Second Edition

A Primer in Quantum Mechanics and Chemical Bonding

The Bell that Rings Light

Other World Scientific Titles by the Author

The Bell that Rings Light: A Primer in Quantum Mechanics and Chemical Bonding
ISBN: 978-981-256-705-5
ISBN: 978-981-256-706-2 (pbk)

Second Edition

A Primer in Quantum Mechanics and Chemical Bonding

The Bell that Rings Light

Joseph J. BelBruno
Dartmouth College, USA

World Scientific

NEW JERSEY • LONDON • SINGAPORE • BEIJING • SHANGHAI • HONG KONG • TAIPEI • CHENNAI • TOKYO

Published by

World Scientific Publishing Co. Pte. Ltd.
5 Toh Tuck Link, Singapore 596224
USA office: 27 Warren Street, Suite 401-402, Hackensack, NJ 07601
UK office: 57 Shelton Street, Covent Garden, London WC2H 9HE

Library of Congress Cataloging-in-Publication Data
Names: BelBruno, Joseph author | Wallace, Dorothy (Dorothy I.) Bell that rings light
Title: A primer in quantum mechanics and chemical bonding : the bell that rings light / Joseph J. BelBruno, Dartmouth College, USA.
Other titles: Bell that rings light
Description: Second edition. | New Jersey : World Scientific, 2026. | Revised edition of: Bell that rings light / Dorothy Wallace & Joseph J. BelBruno. Singapore ; Hackensack, NJ : World Scientific, c2006.
Identifiers: LCCN 2026008594 | ISBN 9789819830688 hardcover | ISBN 9789819831784 paperback | ISBN 9789819830695 ebook for institutions | ISBN 9789819830701 ebook for individuals
Subjects: LCSH: Quantum chemistry--Mathematics | Schrödinger equation | Chemical bonds--Mathematical models
Classification: LCC QD462.6.M39 B45 2026
LC record available at https://lccn.loc.gov/2026008594

British Library Cataloguing-in-Publication Data
A catalogue record for this book is available from the British Library.

For any available supplementary material, please visit
https://www.worldscientific.com/worldscibooks/10.1142/14807#t=suppl

Desk Editor: Shaun Tan Yi Jie

Typeset by Stallion Press
Email: enquiries@stallionpress.com

For Kathy:
Thank you for all of the
wonderful years and your patience
with my never-ending projects.

And thank you to Joe
(@belbruno_hues) for
professional assistance
with the artwork.

Acknowledgement

The author acknowledges the support of the National Science Foundation for the "Dartmouth Mathematics Across the Curriculum Project" which funded the experimental first-year science curriculum that led to the first edition of this text, co-authored with Professor Dorothy Wallace.

Preface

The origin of the first edition was an experimental course entitled "Integrated Mathematics and Physical Sciences" funded by the National Science Foundation and covering Introductory Calculus, Physics and Chemistry for science and engineering majors. The disciplines were interwoven and extended over the entire first year. The first edition combined mathematics and chemistry even though the authors 'spoke different languages' so to speak. In this edition, some of the detailed mathematics is separated into supplementary chapters and the important steps in the mathematics and the final results are carefully integrated into the chemistry from the perspective of a mathematically inclined chemist. The intention is to provide more continuity and a single voice.

The text begins with early experiments that turned the world of physics "upside down" with results that could not be explained by classical or Newtonian physics. The theory is presented using explicit mathematics for model (solvable) problems. Analogies and references to classical theory are used to eventually present the solution to the hydrogen atom problem. We then move to multi-electron atoms and molecules, systems that have no exact quantum mechanical solution. Elementary models are introduced and complexity is added until we arrive at molecular orbital theory for simple molecules. We conclude with some real-world applications to indicate the utility of quantum chemistry.

Most chapters are concluded with a small number of exercises for the reader. These are not necessarily the type of challenging problems that might be encountered in a chemistry midterm. Rather, they are intended to reinforce the concepts presented in the respective chapters. To that end, we have supplied worked solutions to the exercises rather than simple answers. The intention (and hope) is that the presentation will be suitable for those who might just want to learn a little chemistry, but also as a starting point for those who plan to pursue a career as a chemist.

About the Author

Dr. Joseph J. BelBruno is Professor Emeritus of Chemistry at Dartmouth College, USA. He is also Founding President of New Hampshire Nanotechnologies, Co Founder & Scientific Advisor of Allergy Amulet, and Chief Technology Officer of FreshAir Sensor, leading the latter to commercialize molecularly imprinted polymer sensors. This work has led to the development of sensors for biological molecules (amino acids, proteins, and disease markers), environmental contaminants (nicotine, CO, mold, and marijuana), and diagnostic tests. His other research focus is on cluster chemistry. He has published more than 130 papers and holds over 20 patents.

Contents

CHAPTER 1

The Beginning of Atomic Modelling

The notion that the material world is composed of "atoms" is an old one. The first record we have of it is the version held by Democritus, in opposition to Aristotle's world picture. It never dominated the scientific imagination until well after Galileo invented the telescope. Seeing the Milky Way dissolve into a myriad of individual stars did something to the scientific imagination of the Renaissance that has yet to be completely undone. In short, if the universe at large is filled with distinct stars wheeling about, then why shouldn't the universe in miniature behave thus also? And suddenly Democritus's model was no longer about something preposterous.

What you *cannot* do with the simple macroscopic world model, Newtonian physics, is explain chemical reactions. When baking soda mixes with vinegar, the outcome supersedes anything that can be described by an elastic collision between two atoms. Newton's laws are inadequate for describing what happens when you pour a base such as sodium bicarbonate into an acid, for example, vinegar. While physics was enjoying a flowering of ideas in the late Renaissance, chemistry was still all about alchemy rather than what we now distinguish by the name of science. Once the idea of the atom was firmly ensconced in the imagination, however, chemists were able to base a fruitful approach on the simple assumption that *atoms of different substances behaved differently and hence had different structures from one another*. This concept allowed chemists to make successful predictions of chemical reactions.

The previously assumed indivisible atom became divisible enough to share electrons. Protons were massive and captured the essential nature of the atom, such as whether it was a hydrogen atom or an oxygen atom. Electrons were small and could be lost or gained during chemical reactions. In fact, a chemical reaction itself could be defined in terms of electron transfer among atoms. This gain and loss of electrons within an atom didn't much affect the weight of the atom, nor did it change it into another substance. Of course, to keep an atom in a neutral state, that is a chemically inactive state, it is necessary to balance various forces, so protons and electrons should be equal in number in the unreactive atom. The existence of atomic weights requires something additional in the atom, so the massive, electronically neutral neutron must also be present. With this basic equipment in place, chemists proposed rules for different kinds of chemical reactions. These rules were assumptions or empirical observations. The major assumption was that if the universe at large consists of planets revolving around stars, then in miniature, atoms should do the same. This led to a simple planetary-like model. Electrons should revolve around the nucleus of protons and neutrons. The nucleus should never be disturbed. Certain numbers of electrons fit in each orbit. The outside orbits lose or gain electrons first, according to simple mathematical rules with integer labels. Then, it became a matter of working out which electrons were required to predict the observed reactions.

Should it surprise us that the model these scientists postulated worked so remarkably well to explain so many chemical phenomena? In fact, it worked to such an extent that it is still being taught to high school students today, a hundred years later. It had only a few parameters to adjust, namely the number and capacity of electron orbits and rules for losing, gaining and sharing of outside electrons. Yet it explained, in the end, a virtual infinity of phenomena. No wonder that the picture Niels Bohr eventually proposed of an atom, that planetary-like model, remains even today the one that leaps immediately to mind for anybody who finished high school chemistry.

We will look at the series of experiments that led to the current atomic model in later chapters. For now, we will take a cursory look at two results, those of Balmer and Niels Bohr, as illustrations of the progression to the modern world of chemistry. Johann Jakob Balmer found some interesting numerical properties in an experiment that was described in the literature, but that result was not possible to explain without a different model of the atom. The experiment: if you input energy into a cell of hydrogen atoms, the atoms will become "excited". This excitation of the atoms means that they gain energy and the atom eventually decays back into its unexcited "ground" state with the emission of the excitation energy, whose dispersed wavelength distribution is shown in Figure 1.1.

Only certain frequencies were observed. For a given type of atom, the same frequencies were always present. Atoms of different substances each had a characteristic "spectrum", as it was called. Bohr later explained this phenomenon in terms of his own model of the atom by saying that orbits containing electrons occurred at discrete intervals, stationary states, pictured as distances from the nucleus, and that electrons had to "jump" from

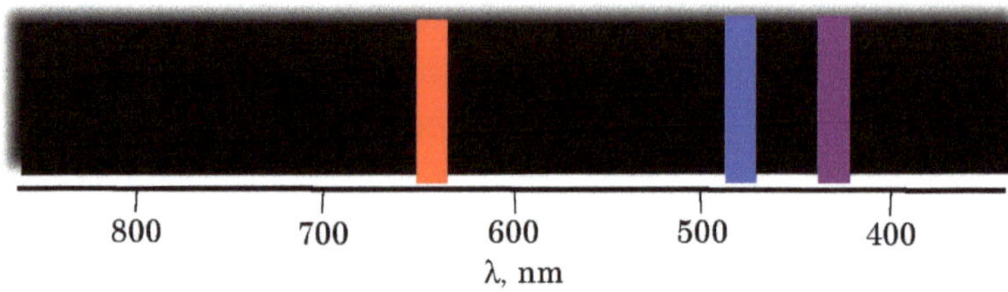

Figure 1.1. The Balmer spectrum of hydrogen. Note that only three wavelengths are present in this spectral range.

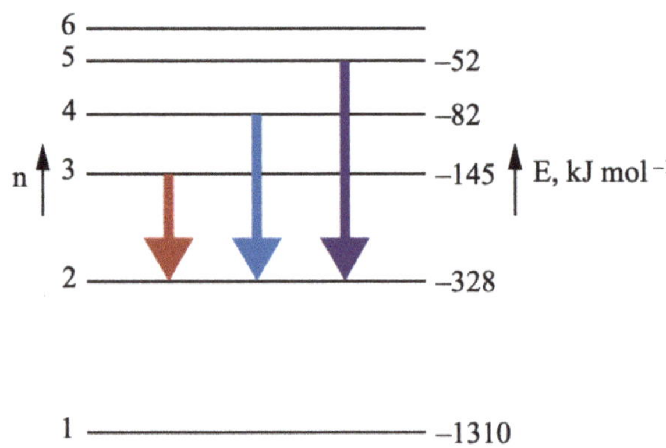

Figure 1.2. The energy level changes within the hydrogen atom resulting in the emission spectrum shown in Figure 1.1.

one of these to another, so that the difference in energy required to promote the electron resulted in a discrete line appearing in the emission spectrum of the atom. In energy level schematic form, this is shown in Figure 1.2.

There were still unresolved problems with the Bohr model. For starters, the underlying assumption that the microscopic universe should mirror the macroscopic one must be set aside. Secondly, this kind of behavior is not something you get from Newton's laws, another of the planetary-like system's underlying premises. And worst of all, without Newton's laws there is no physical mechanism to explain any behavior of any sort that an electron might have. So, we have a model with great predictive value, but one whose incapacity to explain persistent phenomena

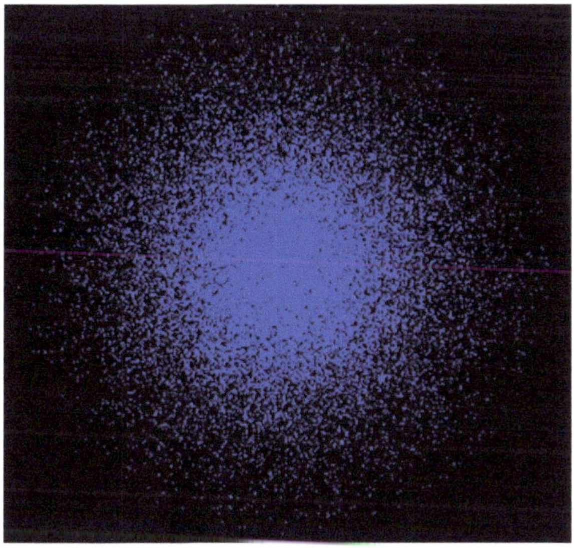

Figure 1.3. Plot of the electron probability density as a function of the radial distance from the nucleus for a hydrogen 1s orbital. The relative density of points at any value of distance is related to the likelihood of localizing the electron near that point.

such as chemical reactions forces the scientific community to rethink the very premises on which the model was initially based.

Later, quantum mechanics would offer a mathematical model of the energy states of the hydrogen atom and it would yield shapes like that shown in Figure 1.3 which, of course, puts an end to this planetary-like atomic model.

Next, we will work through the development of quantum mechanics by examining critical experiments and theoretical stages to reach our modern approach. That approach provides results such as that shown in Figure 1.3.

CHAPTER 2

Some Important Experiments

The two fundamental concepts we need before we approach the idea of quantization are the nature of light and the nature of the electron. You might wonder why we would want to ponder these topics; after all, light is a wave and an electron is a particle. That is clear, or is it? In our macroscopic view of the world, such assumptions would be correct, but we need to look more closely at the microscopic nature of things. The best way to consider these topics is through a brief review of scientific history. Much of what we are about to do depends on our understanding of Coulomb's law, therefore we'll begin with a review of that topic and then look at seminal experiments that defined the nature of light and matter.

Coulomb's law defines the forces between charged objects:

$$F = -C\frac{q_1 q_2}{r^2} \tag{2.1}$$

where q_i is a charge and r is the distance between the charges. The Coulombic potential energy is simply the negative integral of the force:

$$V = -\int F dx = -C\frac{q_1 q_2}{r} \tag{2.2}$$

If Equations (2.1) and (2.2) leave you uncomfortable, your memory might be jogged by looking at a graphical representation of the Coulombic potential energy as in Figure 2.1.

We "assume" that electrons are particles, but why is that? What experimental results led us to this assumption? We can trace history back to Faraday and his electrolysis experiments. In those experiments, an ion "collided with" an electron and deposited metal atoms at an electrode. Faraday's observations included the fact that the mass of deposited material depended upon the current, which is given in units of Coulombs (units of charge) per unit time. Of course, one may relate the total charge to the number of electrons. Therefore, the mass of material (atomic or molecular) deposited at the electrodes was a function of the number of electrons passing through the system and the molar mass of the ion.

Does our consideration of the nature of the electron really matter? Scientists in the late 19th century were concerned with "cathode rays" and their nature — were they really waves? Not according to Faraday's results. The "rays" were actually beams

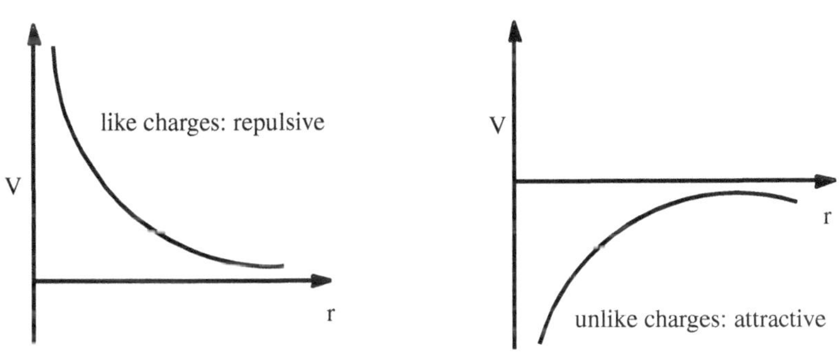

Figure 2.1. Graphical representation of the Coulombic potential as given by Equation (2.2).

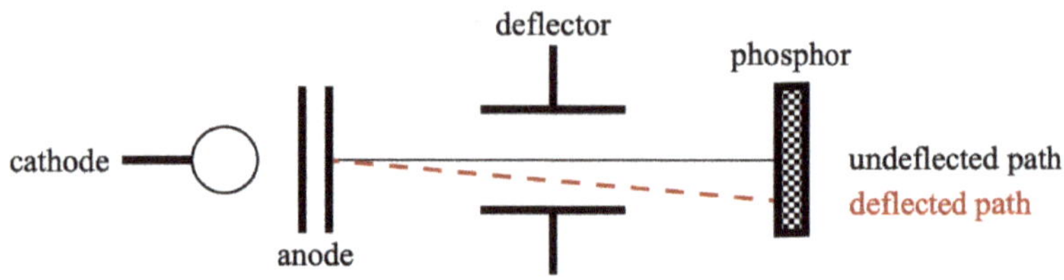

Figure 2.2. The experimental apparatus of J.J. Thomson. Electrons are emitted by the cathode, accelerated by the anode and detected by a phosphor at the end of the tube. A magnetic or electric field may be applied and will deflect the beam of electrons.

of particles, electrons. For more evidence, look to the experiments of J.J. Thomson, as shown in Figure 2.2.

Thomson observed that one could deflect the "rays" with either an electric or a magnetic field — can you do that to a wave? Of course not, that behavior is characteristic of matter. Thomson, from this experiment, found the ratio of the electron charge to its mass, e/m, by analysis of deflecting electric and restoring magnetic fields. The value of e/m was 1,000 times greater than that for any known ion, so the question became: was e large or was m small?

The definitive experiment by R.A. Millikan proved that all electric charges were multiples of 1.6×10^{-19} C (the elementary charge, e). Therefore, he could calculate the mass of the electron. A schematic of his apparatus is shown in Figure 2.3 and now briefly described.

Oil droplets were charged by collisions with ions. In the absence of an additional potential, they fell under the force of gravity. In the presence of an external potential, the droplet acceleration was changed; from this change, m was calculated to be 9.1×10^{-28} g. The particle nature of electrons was now well established: particles may be deflected and tracked. The electron was clearly a particle.

Waves are characterized by frequency ν, wavelength λ, and amplitude. We can relate the frequency and wavelength of light

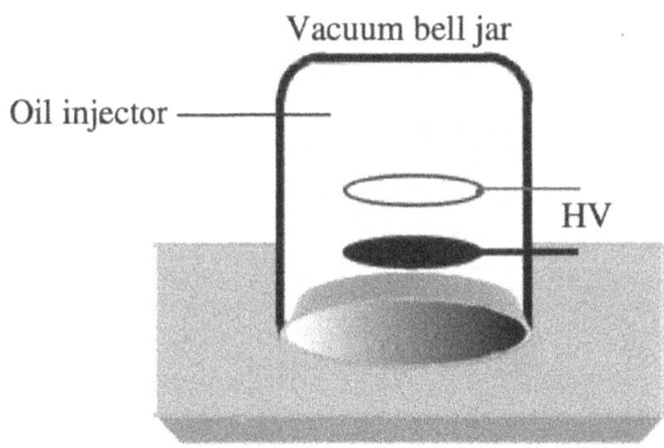

Figure 2.3. Schematic diagram of the Mullikan oil drop experiment. An atomizer injects charged oil droplets into the vacuum chamber where they fall under the influence of gravity. The application of an electric field alters the motion.

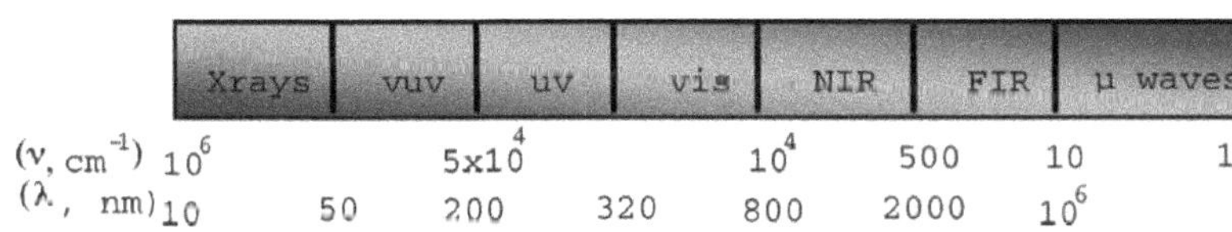

Figure 2.4. The spectrum of electromagnetic radiation with the frequency in wavenumbers or cm^{-1}.

from "wave mechanics":

$$\lambda\nu = c \tag{2.3}$$

where c is the speed of light, $2.9979 \times 10^8\,\mathrm{m\,s^{-1}}$ or approximately $3 \times 10^8\,\mathrm{m\,s^{-1}}$. We represent a portion of the spectrum in Figure 2.4. The frequency, expressed in wavenumber, cm^{-1}, and the wavelength, in nm, increase in opposite directions.

The nature of light was also obvious, at least in the late 1800s — it must be a wave. Let's explore that characterization. We begin by looking at blackbody radiation. We consider the topic qualitatively and use the result of the exact mathematical manipulation.

A blackbody is one that absorbs and emits all frequencies. Experimentally, we accomplish this by the following process. As you heat a body (say, a stove burner), you will observe that the perceived color of the object will make a transition from red to yellow-orange. It could turn white, and even blue, as the temperature increases. The color appears to change because the maximum intensity of the observed light moves to higher frequencies (lower wavelength) as the blackbody is heated. We can record the output, the emission spectrum, of the blackbody and the result is shown in Figure 2.5 for two different temperatures.

The density of radiative energy, $\rho(\nu, T)$, for a frequency between ν and $\nu + d\nu$ (in $\mathrm{J\,m^{-3}}$) is given from classical theory by the following equation, sometimes called the Rayleigh-Jean model:

$$\rho(\nu, T)d\nu = \frac{8\pi kT}{c^3}\nu^2 d\nu \tag{2.4}$$

where c is the speed of light and k is the Boltzmann constant, which has the value $1.38 \times 10^{-23}\,\mathrm{J/K}$. The problem is that the

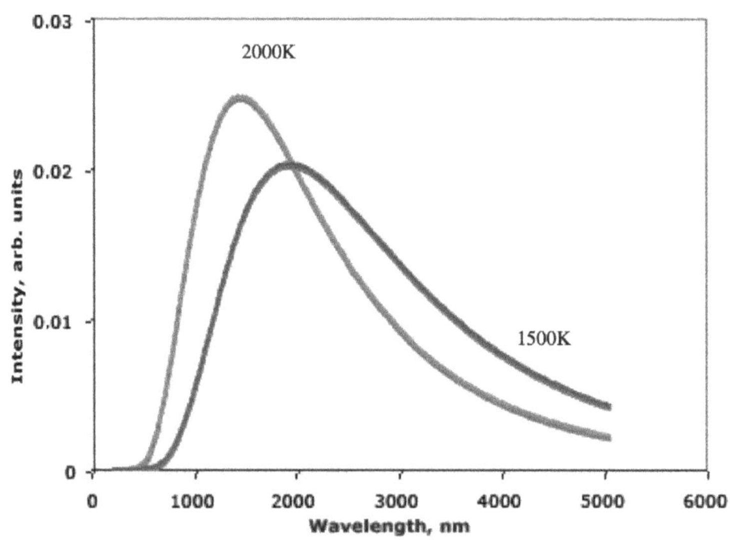

Figure 2.5. Experimental blackbody curves (unnormalized) for two different temperatures.

theory does not agree with the experiment. Theory indicates that the intensity rises as ν^2 (or λ^{-2}) and asymptotically goes to infinity for $\lambda = 0$. This latter result is known as the "ultraviolet catastrophe". Max Planck found the solution to the dilemma. In his model, the radiation is due to oscillations of the electrons in the atoms of the blackbody; this is an antenna problem. In the classical Rayleigh-Jean model, the energies of the oscillators had no restrictions. Classical physics is continuous (all values of the variables are allowed), not discrete (variables are allowed to take on only certain values). Planck *arbitrarily* assumed that the oscillator energies were discrete and restricted to integral multiples, n, of the frequency ν. That is, the energy of the oscillator took values $E = nh\nu$. The energy was *quantized*. Using the techniques of what we now call statistical mechanics, Planck derived an equation (Eq. (2.5)) which agreed with all of the experimental data, when $h = 6.62 \times 10^{-34}$ J s. This is known as Planck's constant:

$$\rho(\nu, T)d\nu = \frac{8\pi h\nu^3}{c^3}\frac{d\nu}{e^{\frac{h\nu}{kT}} - 1} \tag{2.5}$$

Planck's result was not widely accepted in the physics community of the time (early 1900s). The theoreticians were certain that a classical solution to this problem would eventually be discovered. Even Planck was uncomfortable with the quantized results, but this was the beginning of the modern microscopic view of the atomic world. This was the start of **quantum mechanics**.

The final, convincing evidence for the legitimacy of Planck's explanation of the blackbody radiation results came from Einstein and his Nobel Prize-winning theory of the *photoelectric effect*. What is the photoelectric effect? It is schematically shown in Figure 2.6. In short, ultraviolet light causes electrons to be emitted from a metallic surface held in a vacuum. The following *experimental observations* were well known at the time, but were

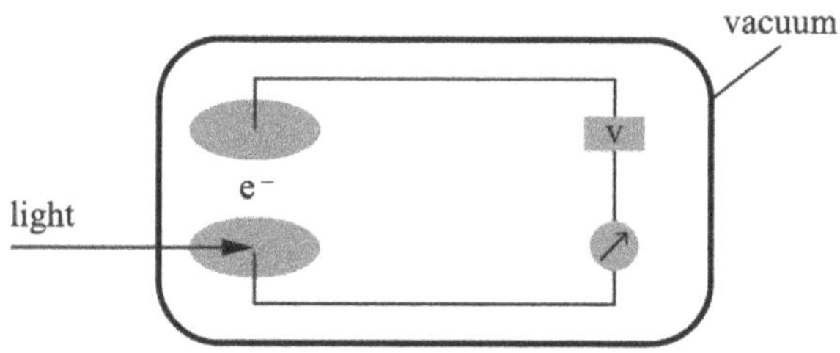

Figure 2.6. The apparatus used to observe the photoelectric effect. Light strikes the photocathode, an electron is emitted, moves toward the anode and is detected.

left unexplained by classical physics:

- the kinetic energy of the emitted electrons is independent of the *intensity* of the incident light;
- there is a threshold frequency and below it, there is no photoelectric effect; and
- the kinetic energy of electrons ejected from the cathode increases with increasing *frequency* of the incident light.

It is instructive to examine the predictions from classical physics. It would be expected, *if classical physics were valid*, that

- as the intensity of the light increases, so does the induced oscillation of the electrons and they break away. That is, higher intensity induces more violent oscillation and higher electron kinetic energy at escape; and
- the photoelectric effect should be frequency independent, but intensity dependent.

Clearly these are not the observations that were reported in the scientific literature and studied by Einstein.

Now, we return to Einstein's work. He knew of and used Planck's results for the oscillators. Einstein proposed that light consisted of packets of energy, $E = h\nu$, called *photons*. Then, using conservation of energy, he wrote

$$\frac{1}{2}mv^2 = h\nu - \phi \tag{2.6}$$

where ϕ is the work function or threshold energy for release of an electron from a particular metal. This is the minimum energy required for electron ejection, $\phi = h\nu_o$. When $h\nu \geq \phi$, an electron is ejected and the excess energy, the energy above ϕ, becomes kinetic energy of the ejected electron. Finally, the number, but not the energy, of the ejected electrons will increase as the number of photons, or the intensity of light, increases. There is exact agreement between the experimental results and Einstein's model. One way to check this agreement is to measure the kinetic energy of the ejected electrons. How do we make these measurements? We use a stopping potential, V_s. That is, the electrons are directed toward a negative electrode and the potential is increased until the electrons are stopped. The equation governing this process is:

$$\frac{1}{2}mv^2 = -eV_S$$

where e is the constant known as the electric charge, 1.6×10^{-19} C. We can plot V_s versus ν or kinetic energy as in Figure 2.7. Now, we have a problem: we have assumed that light could be described as a wave. Einstein just invoked a particle-like property to explain an experiment.

Is light a wave or a particle? And, if light is a particle as well as a wave, what does that say about a particle? Can a particle also be thought of as a wave? In particular, is an electron a wave as well as a particle?

To address this issue, we skip over some important theoretical and experimental advances for the time being and go to 1924. We will not try to derive the solution to this apparent dilemma, but rather note that de Broglie realized that, if light

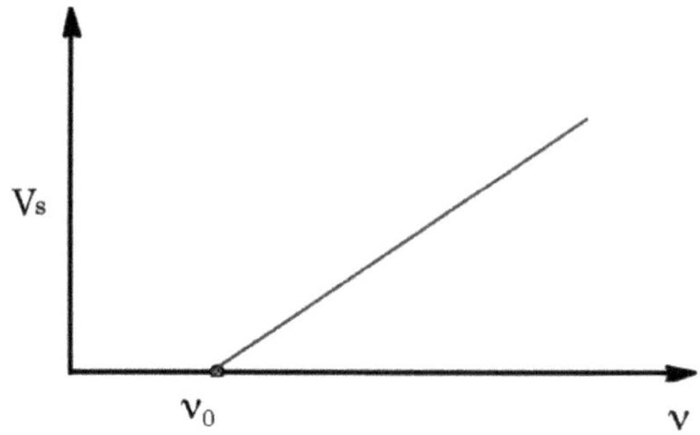

Figure 2.7. Stopping potential for the photoelectron as a function of incident frequency on the cathode.

had the properties of a particle, then it must have momentum. He deduced that the relationship between the wave property of wavelength and the particle property of momentum for a photon (quantized light) could be described by:

$$\rho = \frac{h}{\lambda} \tag{2.7}$$

and we can relate the momentum of a photon to its wavelength. More importantly, de Broglie asserted that this was true for matter, where $p = mv$, as well as for light, where $p = mc$. A practical result of this equality is the electron microscope, the imaging of an object using a beam of electrons. You have probably seen electron micrographs of insects, cells or atoms on a surface. Electron microscopes are capable of imaging objects as small as tens of nanometers. This high resolution results from the small wavelength of the electron traveling at a substantial fraction of the speed of light (large momentum).

We continue our chronological digression by considering the work of Walter Heisenberg, also in the mid-1920s. He carefully considered the way in which the measurement of the position of an object using a photon of a given momentum would affect the position of the object being measured. Consider the following: if we wish to locate an electron to within a distance Δx, we must use light of that wavelength or we won't "see" the electron, at least not to that resolution. If we want to increase the resolution or accuracy, then λ must be smaller than Δx, but p gets larger as the wavelength is reduced and more momentum will be transferred during the collision causing a larger uncertainty in the position of the observed object. The uncertainty relationship is known as the Heisenberg Uncertainty Principle:

$$\Delta x \Delta p \geq \frac{\hbar}{2} \tag{2.8}$$

where "h-bar" is

$$\hbar = \frac{h}{2\pi}$$

This is a fundamental law for the act of measurement at the atomic scale; position and momentum cannot be simultaneously and exactly determined.

Exercises

1. Find the frequency of red light at a wavelength of 650 nm. Find the frequency of 230 nm light. Into which region of the electromagnetic spectrum do the wavelengths fall?
2. Rearrange the Rayleigh-Jean equation in terms of wavelength instead of frequency. Calculate the radiative energy density at 300 K for incident light at wavelengths of 100 nm and 1,000 nm. Can you see the issue with classical theory for blackbody radiation?
3. The threshold energy, ϕ, for emission of an electron from copper metal is 4.7 eV. Will light at 300 nm cause ejection of an electron from the copper surface? Will light at 132 nm cause ejection of the electron?
4. What is the momentum for a 700 nm photon? What is the momentum if the photon has a wavelength of 350 nm?

CHAPTER 3

Early Quantum Mechanics

Now, on to the atom. The atom contains electrons and we have seen that we may influence the energy of those electrons using electromagnetic radiation. Therefore, our discussion of the nature of electrons and light will be significant in the discussion of the atom. Coulomb's law will also be invoked in the upcoming discussion.

Prior to the work of Rutherford, the atom was considered an object of constant density across its volume. J.J. Thomson proposed what was known as the "plum pudding model". He postulated that the atom was a mass of constant positively charged density with electrons scattered about like "raisins in a pudding". This model, which on any level would be incorrect due to Coulomb's law, was disproved experimentally by Rutherford. Rutherford (actually Rutherford and his graduate student, Marsden) scattered α-particles, He^{2+} (helium nuclei stripped of electrons), into metal foils and the data indicated that most of the mass of an atom was localized and the rest of the atom consisted of empty space. These conclusions resulted from an analysis of the scattering of the α-particles from the foils. When the α-particles impacted the nucleus, in which most of the mass was localized, the scattering angle was quite large, nearly 180° or directly back at the source of the α-particles. These backscattering observations were relatively rare, since most α-particles simply passed through the foil, undeflected. The conclusion cited above, that most of the atom is empty space, is obvious from these results. This work led to the familiar "orbital" model of the atom, as shown in Figure 3.1, which also violated the physics of charges. Classical physics tells us that electrons cannot exist in a stable orbit around a positive nucleus. The electron would nearly instantaneously radiate while dropping into the positively charged nucleus. While the new model was sufficient to explain the distribution of mass, it could not explain the reality of the atom, at least not in terms of classical physics. A new kind of physics was required: **quantum mechanics**.

So finally, our history leads us to Niels Bohr and early quantum theory. Bohr's analysis of the problem has errors and limited

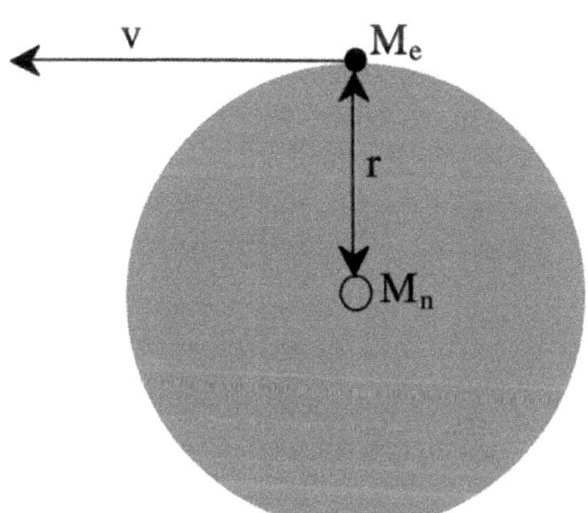

Figure 3.1. The Rutherford representation of the atom.

applicability. However, it is the start of a new field, explains the data available at the time and provides an indication of the new direction physical thought had taken. Therefore, we'll treat it in detail here.

What was the available data? What information set Bohr off on this new path? If we compare the spectrum of the hydrogen atom, the simplest atom, to that of a blackbody, the difference is striking. The blackbody radiates at all wavelengths, while the only visible light emitted by the excited hydrogen atom is at three very narrow wavelengths of red, blue and violet light seen in Figure 1.1. These data were well known when Balmer used simple curve fitting and came up with an equation to fit the frequency data:

$$\nu = 8.2202 \times 10^{14}\left(\frac{1}{4} - \frac{1}{n^2}\right) Hz, \quad n = 3, 4, 5, \ldots \tag{3.1a}$$

We can rewrite this, converting frequency ν to wavenumber $1/\lambda$:

$$\frac{1}{\lambda} = 109{,}680\left(\frac{1}{4} - \frac{1}{n^2}\right)\text{cm}^{-1}, \quad n = 3, 4, 5, \ldots \tag{3.1b}$$

Equation (3.1b) is known as the *Balmer formula*—this equation was later generalized to any integer by Rydberg in what we now call the Rydberg equation, where n_i and n_f are respectively the integers describing the initial and final orbital energy levels:

$$\frac{1}{\lambda} = 109{,}680\left(\frac{1}{n_f^2} - \frac{1}{n_i^2}\right)\text{cm}^{-1}, \quad n = \text{integer} \tag{3.2}$$

The constant, *109,678* cm^{-1} in Equation (3.2), is often given the symbol R_H and is known as the Rydberg constant for hydrogen. This is what Bohr had to work with in his theoretical development, but as you can see, it is an extremely significant empirical result. Now, we go back to the orbital model with which we began this discussion. What do we know from classical physics about an electron orbiting a positive charge, as in Figure 3.1? The electron has *angular momentum* given by mvr. There will be a *centripetal force* toward the nucleus, $F = \frac{mv^2}{r}$, and the electron should radiate and fall into the nucleus! **Bohr simply assumed that the electron didn't radiate and didn't fall into the nucleus and examined the consequences.** He built a theory based on four postulates (only #3 is incorrect):

(1) electrons have allowed energies called stationary states;
(2) electrons in stationary states do not radiate; radiation is due to changes in the stationary state of an electron;
(3) electrons move in circular orbits; and
(4) stationary states have angular momentum in multiples of $h/2\pi$; $mvr = nh/2\pi$.

The centripetal force is Coulombic; after all we are looking at an electron and a collection of protons:

$$\frac{mv^2}{r} = \frac{Ze^2}{r^2} \tag{3.3}$$

where Ze is the nuclear charge and e is the fundamental electron charge.

So,

$$\frac{Ze^2}{r} = mv^2 \tag{3.4}$$

but, from postulate #4,

$$v^2 = \frac{n^2\hbar^2}{m^2r^2}$$

where we defined

$$\hbar = \frac{h}{2\pi} \tag{3.5}$$

and

$$r = \frac{n^2\hbar^2}{Zme^2} \tag{3.6}$$

and is defined as the radius of the Bohr orbit. The radius is quantized, that is, it depends only on an integer n for a given value of atomic number Z:

$$r = 0.530\frac{n^2}{Z}\,\text{Å}, \tag{3.7}$$

where 1 Å $= 10^{-8}$ cm. Now look at the energy of such an atom using Equation (3.4):

$$E = K + V = \frac{1}{2}mv^2 - \frac{Ze^2}{r} = -\frac{1}{2}\frac{Ze^2}{r},$$

$$E = -\frac{1}{2}\frac{Z^2e^4m}{\hbar^2n^2}, \qquad n = 1, 2, 3, \ldots \tag{3.8}$$

where n is a **quantum number**. This is the stationary state energy of an electron in a one-electron atom. As $n \to \infty$, $E \to 0$, the electron escapes from the atom and the atom is *ionized.*

Can we obtain Rydberg's equation from Bohr's model? Yes, we even get R_H, the Rydberg constant in terms of fundamental quantities. We easily recognize the cause of the observed discrete hydrogen atom spectrum using the transitions shown in Figure 3.2.

$$\Delta E = E_f - E_i = \frac{1}{2}\frac{Z^2e^2}{a_0}\left(\frac{1}{n_f^2} - \frac{1}{n_i^2}\right), \qquad n = 1, 2, 3, \ldots$$

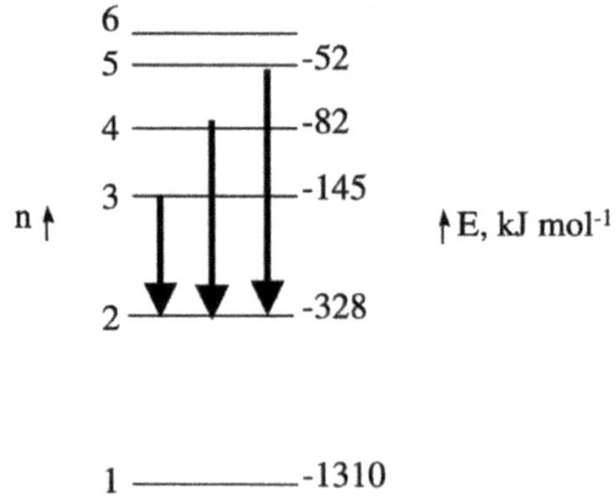

Figure 3.2. Schematic representation of the transitions in the hydrogen atom.

Note that

$$\frac{1}{2}\frac{Z^2e^2}{a_0} = R_H, \quad \text{when } Z = 1 \text{ for hydrogen.}$$

What is wrong with this conceptually simple model? There are two significant problems. We know that Bohr took a leap away from classical physics, but we wish to base the answer to the question on some solid physics. If we know r exactly, this model will also provide us with an exact value of p. This is a violation of Heisenberg's Uncertainty Principle which is fundamental to quantum mechanics and states that some pairs of variables cannot be simultaneously known exactly. One such pair is position and momentum (another is energy and time):

(1) $$r = \frac{a_0 n^2}{Z}, \quad p = \frac{nh}{2\pi} \text{ violates } \Delta p \Delta x \geq \frac{h}{4\pi}$$

(2) Try to apply this model to the helium atom. Now we have two electrons. What do we do? How do we deal with multi-electron systems?

There is no allowance in this model for more than one electron. Some researchers tried to accommodate the extra electrons by using elliptical orbits, but that didn't work. How do we know it failed? We could use the same criterion we used for hydrogen. The question to ask is: Can we reproduce the experimental spectrum? For multi-electron atoms there are far too many lines of nearly the same wavelength; see for example the helium atom spectrum in Figure 3.3. The Bohr model cannot predict the occurrence of these transitions. We need a new model, the "new" quantum mechanics. We shall use a formulation due to Schrödinger (Heisenberg also developed a formulation that is less often applied).

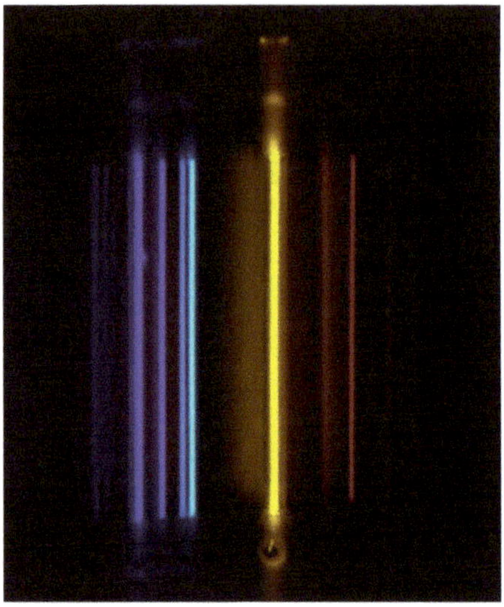

Figure 3.3. Schematic representation of the transitions in the helium atom.

Exercises

1. Why didn't Balmer observe the $n = 6$ to $n = 2$ transition?
2. How does the Rydberg constant vary as the atomic number of the one-electron atom changes?
3. Calculate the wavelength for the $n = 4$ to $n = 2$ transition in the He^+ one-electron ion. Compare this wavelength to that for the hydrogen atom. Can one observe the He^+ transition in the visible region of the electromagnetic spectrum? If not, where in the spectrum would this line appear?
4. We wish to locate a nanoparticle with a diameter of 10 nm. Choose a suitable electron wavelength to image this object. If the wavelength is known to 0.001 nm, what is the uncertainty in the momentum of the nanoparticle?

CHAPTER 4

New Assumptions and a New Model

Certain old assumptions die hard, particularly those having to do with circularity of motion. Bohr's circular orbits would give way to other models. Here we are going to explore not quantum theory itself, but the assumptions upon which it is based.

The prospect of an electron disappearing from one orbit in order to reappear in another was sufficiently disturbing to prompt a review of what an electron might actually be. In fact, the question of what an electron *is* rapidly gave way to the question of what properties it might *have* that one could observe. Presumably it had location, and some locations were more likely than others. Anything that had position could have velocity and acceleration too. And an electron had mass, which meant it also would have momentum. Also, it exhibited energy, and in discrete frequencies, as observed in the Balmer experiment.

This general description means that progress could be made in thinking about the electron by treating it as a wave rather than a particle. Recall de Broglie's duality theorem: we can choose which nature of the electron to invoke depending upon the situation. Classical waves may have any amplitude, but only one frequency. Waves can be superimposed to make new waves, but one can describe what frequencies will be present in the sum. Hydrogen, in its emission of single wavelength red light during the Balmer experiment, always emits the exact same frequency. In this instance, hydrogen, or rather the electron in hydrogen, behaves more like a wave than like a particle.

Now, it is arguably natural to think in terms of observable properties the electron might have rather than actual attributes of the electron itself. In the case of hydrogen, it is very profitable to think of the electron as having not a particular position and momentum, but rather a set of probabilities that describe the likelihood of finding it in a particular region of space at any given time.

A *probability distribution* is a function that describes the likelihood of the electron being in a certain region. This probability is expressed as the area under the graph of the function and above the region in question. By fiat, the total area is one; the function is normalized and the electron must be somewhere. These distributions take the place of what was formerly envisioned as the *location* of the electron. A distribution could equally well describe the average time spent by the objectified electron in a particular region, or some other function related to the energy of the electron.

It may be hard to imagine an electron as something described as a probability distribution and whose average behavior is all one can determine. But one can even go further. One could think of an electron, formerly considered an actual object, not just as an object with characteristic probabilities for being found in various proximities to the nucleus, but as being the actual *probability distribution* itself.

Because this concept of a probability distribution for the electron is critical to the chapters that follow, we will spend

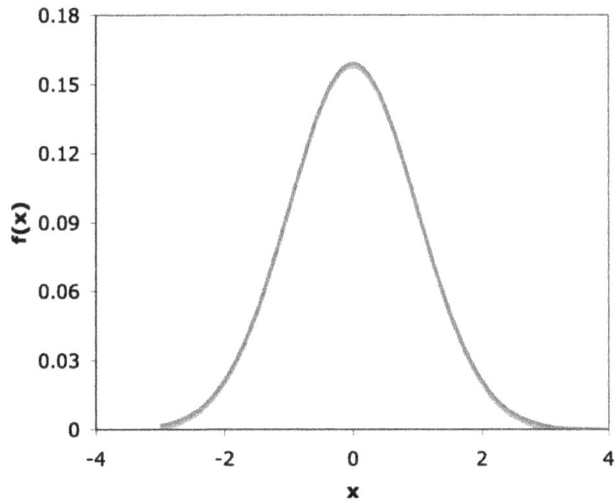

Figure 4.1. The Gaussian distribution function.

a short digression on the topic. Figure 4.1 is a picture of a probability distribution, which you might recognize as the usual bell curve describing grades in a class. If an object moved around on the real axis according to this distribution, it would spend most of its time between $x = -1$ and $x = 1$ and the remainder outside. If you were to reach between two lines drawn at these points and grab it, you would have a chance of catching it there exactly equaling the area under the curve and between the two lines. The total area under the curve is one, which makes the chance of it being somewhere at all also one. This is a very fancy way of asserting the existence of an electron.

These are the new "objects" of quantum physics, replacing the particle nature of the electron in the mind of the quantum mechanist. Furthermore, all the laws will be expressed as if the wavelike properties of the electron were all that mattered. Much has been lost contextually in passing to such a system, especially certain comforting analogies to the macroscopic world. To have an object that can be considered conveniently as wave or particle or probability distribution, depending on the attributes it presents in a given situation, represents an enormous gain in syntax.

To be specific, treating the electron in this manner places it in a class of mathematical objects which can be acted upon by the tools of partial differential equations, functional analysis, abstract algebra and representation theory, which we explore later in optional, supplemental mathematical chapters. The electron behaves not at all like a planet but more like a weight on a spring.

The mathematics brought to bear on the problem is of a depth and flavor attained only in the twentieth century. It represents the combined efforts of both mathematicians and physicists, spawning much of the current activity in both fields. The story of the hydrogen atom in this century exemplifies also what has happened to much of the physical sciences as mathematical "laws", which are theorems, and mathematical "objects", which are sometimes complete abstractions, grew into the conceptual roles formerly filled by familiar physical objects such as planets, billiard balls, and the like.

The Schrödinger equation is a bit of a mystery the first few times it is encountered. And with good reason. It is more or less "given" as something worth studying without any real justification, except perhaps some statement such as: "It is valuable because it works." The Schrödinger equation is not something that can be derived from Newton's laws or Maxwell's equations, or some other accepted "law" of physics. But when you think about it, those "laws" are themselves presented to us and not derived. The justification for them is that they agree with experimental results; in other words, "they work", but somehow the Schrödinger equation seems different, probably because the physics that results from it is so utterly unlike the Newtonian sort to which we have become accustomed.

Having said that one cannot really derive the Schrödinger equation, we now want to make it plausible. There are excellent biographies of Schrödinger and more complete histories of quantum mechanics than that provided here. These provide insight into exactly how it came about. The intent here is to show that it is in some sense reasonable, given what we know of classical physics. Let's try two different conceptual approaches to reaching the Schrödinger equation.

Our first approach will begin by asking a simple question: what is the result of the following difference where x is position and p is momentum?

$$xp - px = ?$$

This is known as a commutation relation and in algebra, of course, the result would be zero. Classical physics would also predict that the result is zero, but in quantum mechanics it is not. We can think about this in the following manner. Classical physics is a macroscopic approximation to quantum physics; in the limit of large dimensions, quantum physics goes over to classical physics. The commutator, defined as $xp - px$, is small, but not zero. The reason that this is true is that in quantum mechanics we need **operators**, denoted by the caret, rather than variables. The result, which we assume rather than derive here, is:

$$\hat{x}\hat{p} - \hat{p}\hat{x} = i\hbar \tag{4.1}$$

where $\hbar$ is Planck's constant divided by 2π and i is the imaginary number, $-1^{1/2}$. In classical physics, the observables, the quantities we measure, are dealt with directly and are functions of time according to Newton's laws. In quantum mechanics, we do not deal directly with the observables, but rather with *operators* corresponding to the observables; $\hat{x}$ and $\hat{p}$ are the position and momentum *operators*, respectively. The result of an operator acting on a function is the observable multiplying the function, that is, the result of the momentum operator acting on some appropriate function is the momentum times that function. If we step back a minute, we see that the commutation relation is actually Heisenberg's Uncertainty Principle in another form. If we assume that the commutation relation is valid, then we can define $\hat{x}$ and $\hat{p}$ from that expression. Define the operator $\hat{x}$ to be "multiply the subsequent function by the coordinate x." Then, in order to satisfy the commutation relationship, $\hat{p}$ must be given by the following:

$$(-i\hbar)\left(\frac{\delta}{\delta x}\right) \tag{4.2}$$

We can test the validity of our assignments using the function Ψ (remember that the operators need a function upon which to act):

$$\begin{aligned}(\hat{x}\hat{p} - \hat{p}\hat{x})\Psi = \hat{x}\hat{p}\Psi - \hat{p}\hat{x}\Psi &= x(-i\hbar)\left(\frac{\delta}{\delta x}\right)\Psi \\ &\quad - (-i\hbar)\left(\frac{\delta}{\delta x}\right)(x\Psi) \\ &= -i\hbar x\left(\frac{\delta}{\delta x}\right)\Psi + (i\hbar)\Psi + i\hbar x\left(\frac{\delta}{\delta x}\right)\Psi \\ &= i\hbar\Psi \end{aligned} \tag{4.3}$$

This indicates that we have correctly defined our operators. If we know $\hat{x}$ and $\hat{p}$ we are all set. Classically we know that there are two energy components, kinetic energy given as $p^2/2m$, and the potential energy V. The total energy is the sum of these two. In one dimension, we may write the total classical energy as:

$$\frac{p^2}{2m} + V(x) = E$$

Now, look at the corresponding quantum mechanical expression using operators:

$$\begin{aligned}\frac{\hat{p}^2}{2m} + V(x) &= E \\ -\frac{\hbar^2}{2m}\left(\frac{\delta^2}{\delta x^2}\right)\Psi + V(x)\Psi &= E\Psi\end{aligned} \tag{4.4}$$

We can define a new operator, the Hamiltonian operator $\hat{H}$. The result of applying this operator on our "eigenfunction" is the total energy of our system:

$$\hat{H} = -\frac{\hbar^2}{2m}\left(\frac{\delta^2}{\delta x^2}\right) + V(x) \tag{4.5}$$

The one-dimensional, time-independent Schrödinger equation is now easily formulated:

$$\hat{H}\Psi = E\Psi \tag{4.6}$$

We can generalize to 3D:

$$\hat{H} = -\frac{\hbar^2}{2m}\left(\frac{\delta^2}{\delta x^2} + \frac{\delta^2}{\delta y^2} + \frac{\delta^2}{\delta z^2}\right) + V(x, y, z) \tag{4.7a}$$

$$\hat{H} = -\frac{\hbar^2}{2m}\nabla^2 + V \tag{4.7b}$$

where the quantity ∇^2 is the standard designation for the partial derivatives in the previous equation. $\hat{H}\Psi = E\Psi$ is an eigenvalue equation, something that appears with regularity in physics and engineering. We've already indicated that the functions Ψ are called eigenfunctions, in this case we also refer to them as wavefunctions, since they will be related to solutions of the wave equation. If we can solve these equations, we are done. How do we solve them? We don't know what these functions called eigenfunctions are and we still have no idea as to the potential energy function. It seems as if we are a long way off from a solution to anything. However, before we try to examine the solution to the Schrödinger equation, let's discuss a different approach to obtaining the equation itself, an approach based upon knowledge of classical wave mechanics.

The wavelike behavior of particles is the fundamental concept upon which the discussion is based. If a particle has attributes of a wave, we can describe it using the classical equation for a traveling wave:

$$\Psi(x,t) = Ce^{i(kx-\omega t)} \tag{4.8}$$

where the wavenumber k is related to the wavelength by

$$k = \frac{2\pi}{\lambda} \tag{4.9}$$

and the angular frequency ω is related to the frequency by

$$\omega = 2\pi\nu \tag{4.10}$$

Of course, all of this begs the question of what it is that $\Psi(x,t)$ represents. If this were a description of water waves, $\Psi(x,t)$ would be the height of the water in the wave above sea level or some other reference height. If this were a description of an electromagnetic wave, then $\Psi(x,t)$ would represent the magnitude of the electric or magnetic field in the wave. But this wave is representing a particle. So what kind of wave is this? The answer, put forth by Max Born, is that $\Psi(x,t)$ is *related to* the probability of finding a particle at a given point in space at a given time. For the present, assume that $\Psi(x,t)$ will represent attributes of the particle that can be measured in the laboratory.

But we know more than the fact that a particle has attributes of a wave. From the work of Einstein and de Broglie, we know that the energy of the particle is related to the frequency of the corresponding wave by:

$$E = h\nu \quad \text{or} \quad E = \hbar\omega$$

and we can solve this for the angular frequency of the wave:

$$\omega = \frac{E}{\hbar}$$

But we also know that the momentum of the particle is related to the wavelength of its associated wave:

$$p = \frac{\hbar}{\lambda} \quad \text{or} \quad p = \hbar k$$

and we can write the wavenumber as

$$k = \frac{p}{\hbar}$$

so that an alternate way of writing the equation for the traveling wave is

$$\Psi(x,t) = Ce^{i(px-Et)} \tag{4.11}$$

It is clear that Ψ has to do with the energy and momentum of the particle that it is representing. Speaking mathematically, how might we extract a measurable value for these quantities from the equation for Ψ that has been proposed? Consider the partial derivative of Ψ with respect to position x:

$$\frac{\delta\Psi}{\delta x} = iCp\, e^{i(px-Et/\hbar)} = ip\Psi \tag{4.12}$$

which may be rewritten as

$$\frac{\delta}{\delta x}\Psi = ip\Psi \tag{4.13}$$

We recognize this to be an eigenvalue equation where the momentum of the particle, p, is the eigenvalue. In short, the mathematical operator

$$(-i\hbar)\left(\frac{\delta}{\delta x}\right) \tag{4.14}$$

extracts the momentum of the particle from the wavefunction Ψ. This may be rewritten as:

$$(-i\hbar)\left(\frac{\delta}{\delta x}\right) \rightarrow p \tag{4.15}$$

In this sense, then, the operator

$$(-i\hbar)\left(\frac{\delta}{\delta x}\right)$$

assumes the role of an analog to momentum for purposes of working with the wave equation. It is sometimes said that this operator "represents" the momentum of the particle. Consider the partial derivative of Ψ with respect to time:

$$\frac{\delta}{\delta t}\Psi = iC\left(\frac{E}{\hbar}\right)e^{i(px-Et/\hbar)}$$

$$\frac{\delta}{\delta t}\Psi = -i\left(\frac{E}{\hbar}\right)\Psi$$

This is another eigenvalue problem, this time with the energy E being the eigenvalue. The operator

$$i\hbar\frac{\delta}{\delta t}\Psi$$

acts to "extract" the energy from the wavefunction, and it may be said to represent energy in quantum theory,

$$i\hbar\frac{\delta}{\delta t}\Psi \rightarrow E$$

In classical physics, the energy and momentum of a particle are related quantities. In particular,

$$\frac{p^2}{2m} + V(x) = E$$

We are now going to use the quantum mechanical operator analogs to energy and momentum in this classical relationship to see how these operators might be related. What justifies doing this? Analogy.

Remember that this is not a derivation, it is a discussion to make the Schrödinger equation seem reasonable. Notice that the classical expression for total energy involves the square of the momentum. In the case of an operator we will interpret this as the operator acting on itself like this:

$$\hat{p}^2 \rightarrow (i\hbar)\left(\frac{\delta}{\delta x}\right) \cdot (i\hbar)\left(\frac{\delta}{\delta x}\right) \tag{4.16}$$

and carrying out the indicated operations

$$\hat{p}^2 \rightarrow \hbar^2\left(\frac{\delta^2}{\delta x^2}\right)$$

Substituting this result into the equation for the total energy,

$$-\frac{\hbar^2}{2m}\left(\frac{\delta^2}{\delta x^2}\right) + V(x) = E \tag{4.17}$$

or

$$i\hbar\frac{\delta}{\delta t}\Psi = -\left[\left(\frac{\hbar^2}{2m}\right)\left(\frac{\delta^2}{\delta x^2}\right) + V\right]\Psi \tag{4.18}$$

which is the Schrödinger equation. The mathematical treatment of classical waves is presented in more detail in Chapter S1 and the interested reader is referred to that chapter for the details.

Exercises

1. Discuss the relationship between the commutator of position and momentum and the Heisenberg Uncertainty Principle.
2. Show that the partial derivative $\frac{\partial}{\partial x}\Psi = ip\Psi$.
3. Show that the partial derivative $\frac{\partial}{\partial t}\Psi = -i\left(\frac{E}{\hbar}\right)\Psi$.
4. Show that $\hat{p}^2 \rightarrow \hbar^2\left(\frac{\delta^2}{\delta x^2}\right)$.

CHAPTER 5

Model Problems in Quantum Mechanics

Before going on to something as complex as an atom, let's look at model quantum mechanical problems in some detail. The customary first problem is the one-dimensional particle in a box. This turns out to be an excellent conceptual model for conjugated dye molecules (see Chapter 13), and a physics model for trapped charged particles. The potential term is shown graphically and mathematically in Figure 5.1.

The formulation of the problem begins with the one-dimensional, time-independent Schrödinger equation. Note that the Schrödinger equation is always valid; the nature of the problem will change the potential energy function, $V(x)$, in that equation:

$$\hat{H} = -\frac{\hbar^2}{2m}\frac{\delta^2}{\delta x^2} + V(x) \rightarrow -\left(\frac{\hbar^2}{2m}\right)\left(\frac{\delta^2}{\delta x^2}\right)\Psi = E\Psi$$

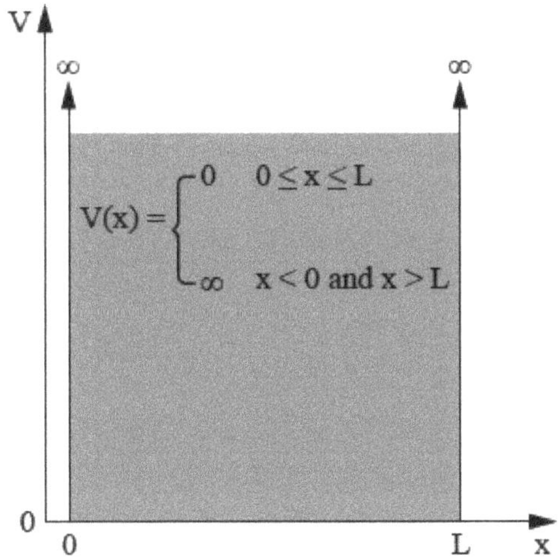

Figure 5.1. The potential energy for the one-dimensional particle in a box.

For this problem, the potential energy is zero within the box and the walls are infinitely high since the potential energy is infinite outside of the box. We expect that the particle will only be found within the box and that the probability (and wavefunction) outside of the box is zero. We are solving for the case where the potential energy $V(x) = 0$. What happens at the walls? This is a boundary value problem, and we solve the equation by using the boundary conditions. In this case, we assume that the wavefunction also vanishes at the walls, $\Psi(0) = 0$ and $\Psi(L) = 0$, where L is the box length.

The general solution to this one-dimensional Schrödinger equation is well known. You could easily solve it with knowledge of differential equations or locate the general solution in tables of solutions to differential equations. The general solution is:

$$\Psi(x) = C\, cos\, kx + D\, sin\, kx, \quad \text{where } k = \left(\frac{2mE}{\hbar^2}\right)^{1/2} \tag{5.1}$$

Now, use the boundary conditions to obtain the specific solution from the general one. At $x = 0$, $sin\ x = 0$ and then C must also equal zero and we do not learn anything about our system. At $x = L$, we must satisfy the remaining boundary equation:

$$\psi(L) = D\ sin\ kL = 0, \quad \text{where } sin\ kL = 0 \text{ when } kL = n\pi,$$
$$n = 1, 2, 3, \ldots$$

The solution is

$$k = \left(\frac{n\pi}{L}\right) \text{ when } kL = n\pi \quad \text{and} \quad n = 1,\ 2,\ 3.$$

and the energy is quantized, where n is the quantum number! Note that nowhere did we assume that the energy was quantized, a'la Bohr. We formulated the problem, solved it and the quantization fell out of the solution.

One last problem remains to be solved – what is D? We will see that this variable, known as the *normalization constant*, ensures that the total probability is equal to 1. We need to define a new quantity. $|\psi|^2$ is known as the probability density, that is, the *probability per unit distance* for a one-dimensional problem. If we integrate the probability density over all space, from $x = 0$ to $x = L$, the result must be unity, since the particle must be somewhere in the box! Let's do the integral using the wavefunction with the as yet undefined value of D:

$$1 = \int \Psi * \Psi dx = D^2 \int_0^L sin^2\left(\frac{n\pi x}{L}\right) dx = \frac{1}{2} LD^2 \qquad (5.2)$$

So, $D = \left(\frac{2}{L}\right)^{1/2}$ and we are done! The results for the wavefunction and the energy are summarized below. These are specific to the problem at hand. Other quantum mechanical problems will have different solutions, but the technique of reaching that solution will be analogous to that we have just completed:

$$E = \frac{n^2\pi^2\hbar^2}{2mL^2}, \quad n = 1, 2, 3 \ldots.$$

$$\Psi_n(x) = \left(\frac{2}{L}\right)^{1/2} sin\left(\frac{n\pi x}{L}\right)$$

We graphically observe the wavefunctions and the probability densities as we vary the quantum number n in Figure 5.2. There is no reason to expect that the wavefunctions will be the same for each value of n. In fact, the solutions look as shown in the figure. The wavefunctions may be distinguished by the number of nodes. A node is defined as a point at which Ψ (and $|\Psi|^2$) are zero. We will not find the particle at this position. Note that the number of nodes increases by one as we go to each successive energy level. This is a general statement regardless of the quantum problem at hand. The wavefunctions ψ are identical to the standing waves generated by a vibrating string, an example with which we are much more familiar.

The energy of the particle in a box depends upon the quantum number, increasing as n increases, see Figure 5.3. However, it also depends upon the length of the box. As L gets very large, the energy decreases, as does the separation between two adjacent energy levels. In fact, in the limit of very large L, the particle in a box asymptotically approaches a free particle. In this limit, we make a transition from quantum mechanics with discrete energy levels to classical physics with continuous energy levels.

What is ΔE for the particle in a box? That is, what is the separation between adjacent energy levels? We need to ask this question because experimentally, the energy differences are what we observe. Recall Bohr's postulate stating that electrons in atoms do not radiate; the radiation is due to changes in stationary states. So, if we observe radiation from an atom, we are

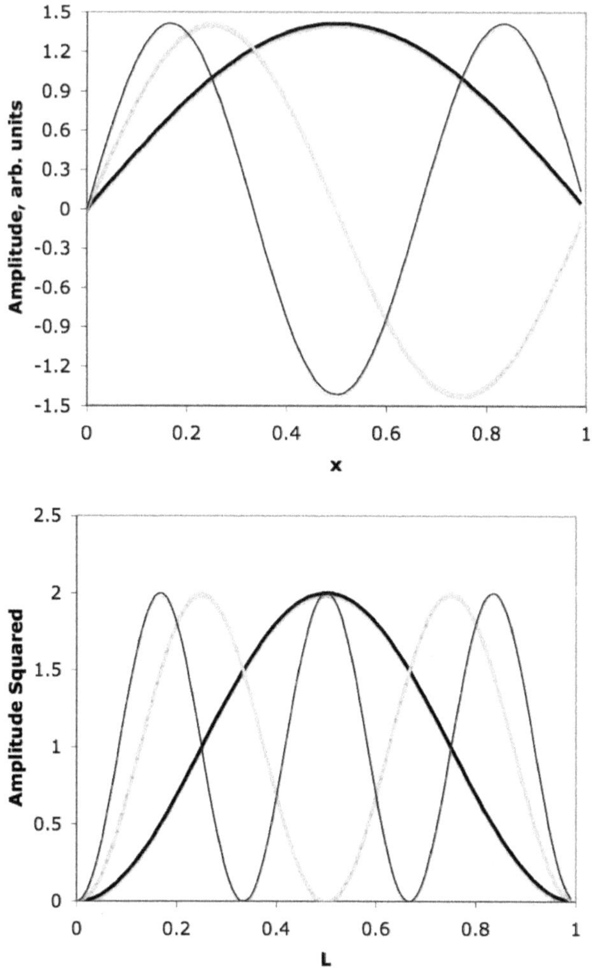

Figure 5.2. The wavefunctions (top) and probability densities (bottom) for the first three states of the particle in a box: $n = 1$ (black), $n = 2$ (gray) and $n = 3$ (green).

observing the energy difference between two levels. The energy difference is readily obtained by use of n and $n + 1$ as quantum numbers in the expression we derived for the one-dimensional case.

$$\Delta E = E_{n+1} - E_n = \frac{(2n+1)\pi^2\hbar^2}{2mL^2} \tag{5.3}$$

Figure 5.3. Schematic representation of the energy levels for the one-dimensional particle in a box problem.

Notice in Figure 5.3 that, unlike the Bohr atom (and all real atoms and molecules), the energy levels get farther apart with increasing n.

An interesting question: what happens if the box is three-dimensional (with infinitely high walls and zero potential in the box)? This provides us with another, but a related, problem to solve. The Hamiltonian operator for a three-dimensional (cubic) box with infinite walls is straightforward and is written below:

$$\widehat{H} = -\frac{\hbar^2}{2m}\frac{\delta^2}{\delta x^2} - \frac{\hbar^2}{2m}\frac{\delta^2}{\delta y^2} - \frac{\hbar^2}{2m}\frac{\delta^2}{\delta z^2} + V(x, y, z),$$
$$\text{where } V(x, y, z) = 0$$

Motion in any direction in the box is assumed to be independent of the other directions, so we have separated the variables and have three separate, analogous equations to solve. The total energy is the sum of the energy in each direction:

$$E = \frac{n_x^2\pi^2\hbar^2}{2mL^2} + \frac{n_y^2\pi^2\hbar^2}{2mL^2} + \frac{n_z^2\pi^2\hbar^2}{2mL^2}, \quad n_i = 1, 2, 3 \ldots. \tag{5.4}$$

The lowest (or ground) energy state is when $n_x = n_y = n_z = 1$ or

$$E = \frac{3\pi^2\hbar^2}{2mL^2}$$

The wavefunction is also related to the one-dimensional case:

$$\Psi_n(x,y,z) = \left(\frac{8}{L^3}\right)^{1/2} sin\left(\frac{n_x \pi x}{L}\right) sin\left(\frac{n_y \pi y}{L}\right) sin\left(\frac{n_z \pi z}{L}\right) \tag{5.5}$$

Both the energy expression and the wavefunction look somewhat different if the box is not cubic, i.e., $L_x \neq L_y \neq L_z$. Exercise 5 will lead you to the correct expression in this instance. Something new also occurs when we look for the next lowest energy or the first excited state where one quantum number is equal to two and the remaining two quantum numbers are still equal to one. This is an effect in quantum energy levels known as "degeneracy" – more than one combination of quantum numbers leads to the same total energy.

We look at one last model problem. This is the quantum mechanical treatment of the one-dimensional harmonic oscillator. Recall that the potential energy is given as $V(x) = \frac{1}{2}kx^2$, where k is the force constant. The Schrödinger equation for this problem is given in Equation (5.6):

$$\widehat{H}\Psi = -\frac{\hbar^2}{2m}\frac{\delta^2}{\delta x^2}\Psi + \frac{1}{2}kx^2\Psi = E\Psi \tag{5.6}$$

Solving this equation is more complicated than for the particle in a box and we only state the solution and write the resulting energy and wavefunctions. The energy is

$$E = \left(v + \frac{1}{2}\right) h\nu \quad \text{where } v = 0, 1, 2 \ldots.$$

Note that the quantum number v can take on all integer values, including zero. This is because the total energy when $v = 0$ is not zero, but $\frac{1}{2}h\nu$. The vibrational energy includes a "zero-point" energy of $\frac{1}{2}h\nu$ and allowing $v = 0$ will not violate the Uncertainty Principle. The zero-point energy is the vibrational energy when $T = 0\,K$. The wavefunctions look like those we have already seen and are in Figure 5.4. The wavefunctions are

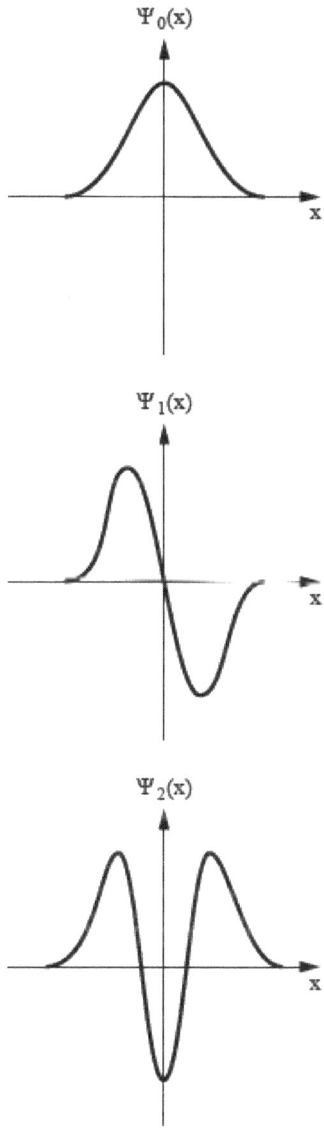

Figure 5.4. The lowest three wavefunctions for a one-dimensional quantum mechanical oscillator.

mathematically $\Psi_v = e^{\frac{-\alpha x^2}{2}} P(x)$, where $P(x)$ is a polynomial expression. Note that the number of nodes also increases by one as we traverse successive energy levels.

Exercises

1. Show that $k = \left(\frac{n\pi}{L}\right)$ and $E = \frac{n^2\pi^2\hbar^2}{2mL^2}$.
2. Use a table of integrals to show that $D = \left(\frac{2}{L}\right)^{1/2}$.
3. Beginning with explicit expressions for the energy levels, show that Equation (5.3) is correct.
4. Calculate the energy spacing between the ground state and the second excited state for a particle in a cubic three-dimensional box.
5. Find the ground state energy for a particle in a three-dimensional box with infinite walls where $L_x \neq L_y \neq L_z$.

CHAPTER 6

The Schrödinger Equation for a One-Electron System

The historical development of the Schrödinger equation is available for your edification in quantum mechanics texts. That development is beyond what we hope to accomplish here. We treat quantum mechanics as axiomatic. The basic underlying assumption that an electron might obey a wave equation is an *axiom* upon which quantum mechanics rests. The alternative assumption of an electron behaving like a body orbiting another body according to Newton's laws cannot lead to the observations with which we began our study in Chapter 1. Unlike Euclidean geometry or Newtonian physics, the answer to be explained was far from believable to the physicists of the era. The physicists of the early part of the 20th century were attempting to put a foundation under atomic phenomena, whose foundation itself undermined Newton. In fact, the axioms themselves were very close to certain of the phenomena. So, while one might use them to predict what would happen, they were very unsatisfying answers to the question *why*?

Certain clues and pointers did lie in the wavelike nature of light, however. After a long and somewhat convoluted development, it was Schrödinger who finally suggested that a particle with no forces acting on it should naturally obey the wave equation. For this he got the Nobel prize. In fact, in the context of particles, the wave equation is actually called the Schrödinger equation.

Of course, an electron that is part of an atom does indeed have forces acting on it, otherwise it couldn't remain attached to its positive core. This is where Schrödinger's equation becomes interesting. There are various candidates for the potential field that would keep an electron in what we will call an "orbit", but we know that it would have to have spherical symmetry. That is, any force acting on the electron should be expressed as a function of r only, where r is the radius from the electron to the center of the atom. In addition, that force should die away at great distance from the core of the atom, because if an electron gets far enough from the positive charge it is lost (the atom is ionized). The simplest candidate for a function of r that does this is $1/r$, and indeed, this function (up to a constant multiple) is correct. In keeping with the spirit of forces acting on springs, we add it to the list of operators affecting our position function for the electron. For reasons of tradition, that function is always called Ψ. In this case, it would be $\Psi(x, y, z, t)$. We will proceed with the time-independent Schrödinger equation for simplicity. Our goal is to recast the Schrödinger equation in a more useful form and we restrict our consideration in this chapter to an atom or ion with a single electron, the hydrogen atom or a hydrogen-like ion.

We have seen that the wave equation in one spatial variable (dimension) has the operator $\frac{\delta^2}{\delta x^2}$. The operator in three dimensions must be unable to distinguish different directions from the origin; the force is independent of the directional coordinate. We saw the three-dimensional Hamiltonian in Chapter 5:

$$\hat{H} = -\frac{\hbar^2}{2m}\frac{\delta^2}{\delta x^2} - \frac{\hbar^2}{2m}\frac{\delta^2}{\delta y^2} - \frac{\hbar^2}{2m}\frac{\delta^2}{\delta z^2} + V(x, y, z)$$

The potential energy is electrostatic, $V = \frac{Ze^2}{r}$, where r is the distance of the electron from the center of the atom in spherical polar coordinates, Z is the atomic number (total positive charge) and e is the fundamental electronic charge.

We can write the Hamiltonian, excluding the potential function, in a shorthand notation using:

$$\nabla^2 = -\frac{\hbar^2}{2m}\left(\frac{\delta^2}{\delta x^2} + \frac{\delta^2}{\delta y^2} + \frac{\delta^2}{\delta z^2}\right) \tag{6.1}$$

so that

$$\hat{H} = \nabla^2 + \frac{Ze^2}{r} \quad \text{and} \quad r = \sqrt{x^2 + y^2 + z^2}$$

The first step is to make a coordinate system change. This change is typically a problem encountered in a calculus course. We will begin by assigning spherical polar coordinate variables for Cartesian x, y, z (see, also, Figure 6.1):

$$\begin{aligned} x &= r\, sin\theta\, cos\phi \\ y &= r\, sin\theta\, sin\phi \\ z &= r\, cos\theta \end{aligned} \tag{6.2}$$

and substitute these identities in the Cartesian Hamiltonian.

It is an interesting exercise, but beyond the scope of this text, to recast the Hamiltonian in terms of these new variables. Here,

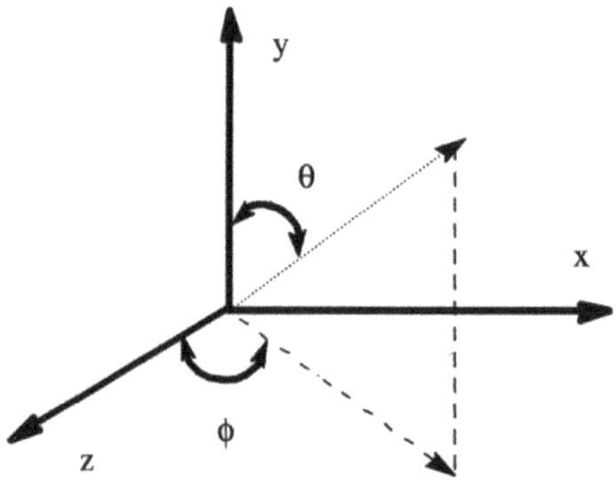

Figure 6.1. Parameters for conversion from Cartesian to spherical polar coordinates.

we move directly to the result of that exercise:

$$\nabla^2 = \frac{1}{r^2}\frac{\delta}{\delta r}r^2\frac{\delta}{\delta r} + \frac{1}{r^2 sin\theta}\frac{\delta}{\delta\theta}sin\theta\frac{\delta}{\delta\theta} + \frac{1}{r^2 sin^2\theta}\frac{\delta^2}{\delta\phi^2}$$

$$\nabla^2 = \frac{1}{r^2}\frac{\delta}{\delta r}r^2\frac{\delta}{\delta r} + \frac{1}{r^2}\left(\frac{1}{sin\theta}\frac{\delta}{\delta\theta}sin\theta\frac{\delta}{\delta\theta} + \frac{1}{sin^2\theta}\frac{\delta^2}{\delta\phi^2}\right) \tag{6.3}$$

We will want to separate variables in the same way we did for the wave equation in Chapter 5, but which variables shall we use? The key here lies in the spherical symmetry. If you set $r = 1$ and let θ and ϕ change, you get points that lie on the sphere of radius one. These are the natural coordinates for points on a sphere. We can set a partial separation of variables into spatial and angular components:

$$\Psi(x, y, z) \rightarrow \Psi(r, \theta, \phi) \rightarrow R(r)Y(\theta, \phi) \tag{6.4}$$

We will have an ordinary differential equation in r, because this variable occurs alone and we will have a partial differential equation in θ and ϕ because we did not further separate these two coordinates.

We can now write the Schrödinger equation for any system. We would need to include the positions of all electrons and the much more complicated potential energy function. The problem is: how can we solve it for any system, or for even one system? It turns out that we can solve the equation exactly only for a one-electron system. All other cases will require some form of approximation. We'll not try to reproduce that one-electron solution here. For now we shall concentrate only on the solutions and interpret them. Recall that Bohr assumed quantum numbers; they result from the solution of the Schrödinger equation. Let's look at the quantum numbers resulting from that solution and assign physical significance to them.

Recall that in our one-dimensional particle in a box we had one quantum number. In the three-dimensional system, we had three. In short, we have one quantum number per coordinate (actually, per squared coordinate or degree of freedom). Below, we look at the quantum numbers and assess their significance.

(1) The quantum number n is the *principal quantum number*, and is related to the energy of an electron in an atom:

$$E = -\frac{me^4Z^2}{n^2\hbar^2} \quad n = 1, 2, 3, \ldots \tag{6.5}$$

The energy level spacing grows smaller as n increases. As $n \to \infty$, $E = 0$ and the atom is ionized. You can easily see that Schrödinger and Bohr were in agreement for the hydrogen atom energy, as they should be.

(2) The *angular momentum quantum number*, ℓ, is required to accurately express the *average* electron distance from the nucleus:

$$r = \frac{a_0 n^2}{Z}\left(\frac{3}{2} - \frac{\ell(\ell+1)}{2n^2}\right), \quad 0 < \ell < n-1 \tag{6.6}$$

$\ell = 0 \Rightarrow s$ orbital

$\ell = 1 \Rightarrow p$ orbital, etc.

(3) The *magnetic quantum number*, m_l, determines orientation in, and is only important for, atoms in a magnetic field. The values are restricted to

$$-\ell \leq m_l \leq \ell$$

There is a fourth quantum number that is necessary, but does not result from the solution to the Schrödinger equation as we have written it. Rather, it results from a relativistic form of the equation. This is the *spin quantum number*, m_s, which is needed for many-electron atoms and has values:

$$m_s = \pm\frac{1}{2}$$

How do we use these quantum numbers? We can specify the state of an atom (and later, a molecule) using quantum numbers. Let's see how by finding the answer to the following question: what are the quantum numbers for the ground (lowest energy) state of the hydrogen atom? We come to the solution by writing the lowest possible values for each of the quantum numbers. That would be the following set:

$$n = 1$$

$$\ell = 0$$

$$m_\ell = 0$$

$$m_s = \pm\frac{1}{2}$$

Now take this concept one step further: what about the first excited state of the hydrogen atom?

$$n = 2$$

$$\ell = 0 \quad \text{or} \quad \ell = 1$$

$$\swarrow \quad \searrow$$

$$m_\ell = 0 \quad m_\ell = 0 \pm 1$$

$$m_s = \pm\frac{1}{2}$$

These results lead us to eight possible sets of quantum numbers:

$$2, 0, 0, \pm\frac{1}{2}$$

$$2, 1, (-1, 0, 1), \pm\frac{1}{2}$$

A table helps to make this clearer and allows a further extension of the concept for hydrogen. There are always $2n^2$ possible combinations of quantum numbers. We divide these into orbitals. Orbitals are maps of the probability of the electron being located in a certain region in space. Orbitals are designated by their angular momentum quantum numbers. The values of magnetic and spin quantum numbers define the electrons within an orbital.

n	ℓ	m_ℓ	m_s	E, kJ mol^{-1}	Number states
1	0 (1s)	0	±1/2	−1310	2
2	0 (2s)	0	±1/2	−328	8
	1 (2p)	−1, 0, 1	±1/2		
3	0 (3s)	0	±1/2	−145	18
	1 (3p)	−1, 0, 1	±1/2		
	2 (3d)	−2, −1, 0, 1, 0	±1/2		

As a result of writing the Schrödinger equation in spherical polar coordinates, we can write a radial part and an angular part for the wavefunction Ψ. As an example, let's look at the wavefunction for the 1s and $2p_z$ orbitals in the hydrogen atom These orbitals have the quantum numbers $n = 1$, $\ell = 0$, $m_\ell = 0$

and $n = 2$, $\ell = 1$, $m_\ell = 0$, respectively.

$$\Psi_{1s} = \left(\frac{1}{4\pi}\right)^{1/2} * 2\left(\frac{1}{a}\right)^{1/2} e^{-r/a} \tag{6.7}$$

$$\Psi_{2p_z} = \left(\frac{31}{4\pi}\right)^{1/2} cos\theta * \frac{1}{\sqrt{3}}\left(\frac{1}{2a}\right)^{3/2}\left(\frac{r}{a}\right) e^{-r/a} \tag{6.8}$$

where $a = \frac{a_0}{Z}$ (for hydrogen, $a = a_0$, the Bohr radius $= 0.529 \times 10^{-10}$ m).

Consider $\Psi^2_{1s(r)}$, the probability per unit volume within the 1*s* orbital, shown in Figure 6.2. There is a maximum at the nucleus. This is correct, but is contrary to our preconceived notion that the electron is not in the nucleus. Our common sense is correct. We are looking at the *probability density* and we need to multiply the probability density (probability per unit volume) by a volume element to get a *probability*. The volume element of the nucleus is very small and so the probability of locating the electron at the nucleus is negligible (but not zero!). It is easier to consider the probability of finding the electron at *r*, regardless of the angular part of the wavefunction using the

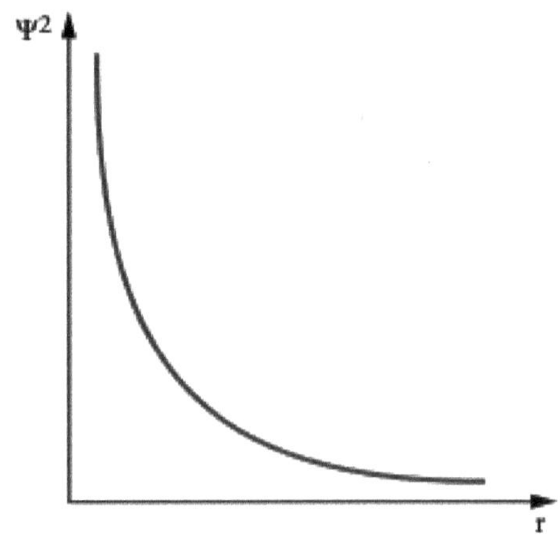

Figure 6.2. The probability density for the 1*s* orbital of hydrogen.

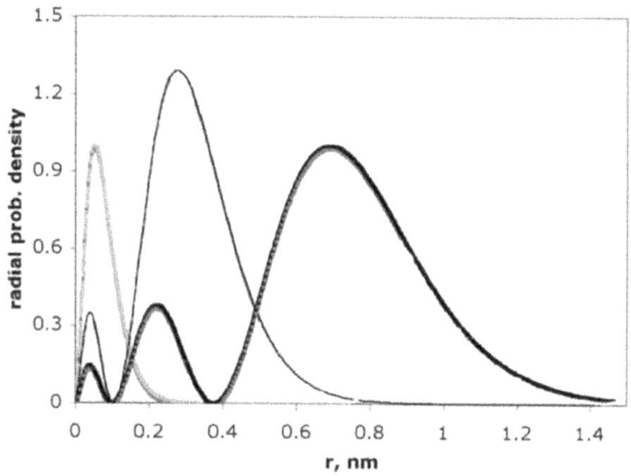

Figure 6.3. The radial probability distribution function for hydrogen s orbitals: $1s$ (gray), $2s$ (black) and $3s$ (heavy black).

separation of variables. So, we construct the *radial probability distribution function*, obtained by integrating over all angles. The result is $r^2R^2(r)$. This is shown in Figure 6.3 for several different s orbitals.

In general, all s orbitals have some small probability near the nucleus. Why do p and s orbitals look different? This seems to have no relationship to the radial function we just discussed. The difference in shape is due to the angular terms. Let's ignore the radial part of the wavefunction and explore the angular term, $\Phi(\theta, \phi)$. In that case, we can write:

$$|\Phi_{2p_z}|^2 \propto cos^2\theta \quad \text{and} \quad |\Phi_{1s}|^2 \propto 1$$

Clearly, $|\Phi_{1s}|^2$ will not have any angular dependence. That is, it will generate a sphere in three dimensions. $|\Phi_{2p_z}|^2$, on the other hand, is a periodic function and the probability density is shown in Figure 6.4. Values of θ range from 0 to π and the angular dependence of the wavefunction (think about rotating the curve in Figure 6.4) will have the well-known "dumbbell" shape. The shapes of these two orbitals are shown in Figure 6.5.

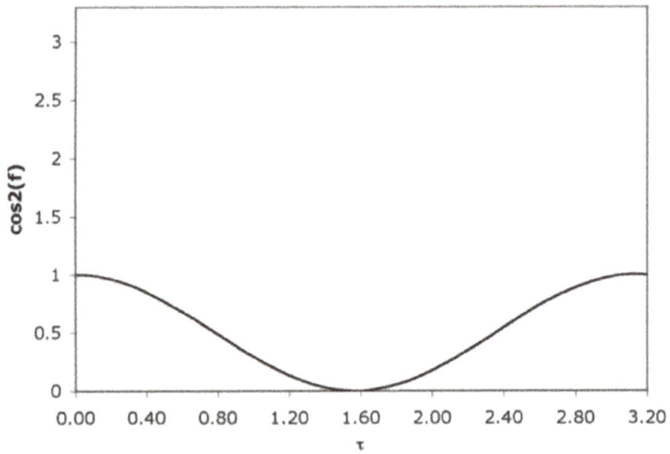

Figure 6.4. The angular dependence (in radians) of the unnormalized hydrogen $2p_z$ wavefunction.

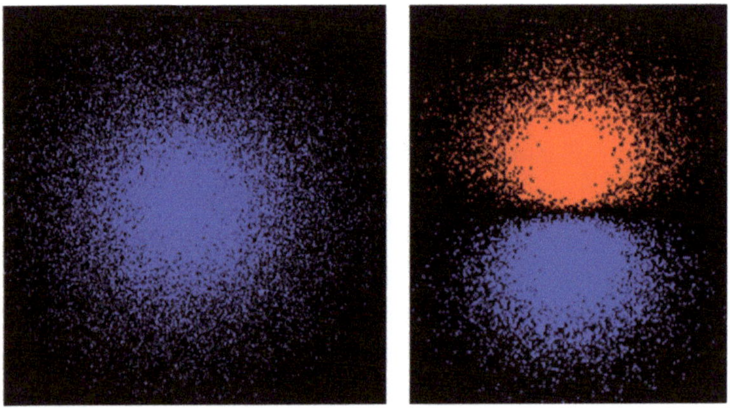

Figure 6.5. Maps of the $1s$ (left) and $2p_z$ orbitals (right).

In this chapter, we saw that we could treat the electron as a probability density. That is, the probability density could be employed to calculate the properties of the electron in the hydrogen atom. We can consider a summary of what type of information we might glean from different aspects of the one-electron probability density:

What are the differences among ψ, ψ^2, R^2, etc.?

- $\psi^2(r, \theta, \varphi) \Rightarrow$ the probability density at r, θ, φ (the probability per unit volume of finding the electron in a volume element near (r, θ, φ)).
- R^2 is the radial probability density. This is the probability of finding an electron between r and $r + dr$, regardless of θ and φ.
- The radial probability is $R^2(r) \times$ volume of a spherical shell between r and $r + dr$ and is $\sim r^2\ dr$, as shown in Figure 6.6. So, the radial probability equals $r^2 R(r^2)\ dr$.

We also saw that we could use the angular probability densities to map out probability regions in space where the electron might be found, providing us with what chemists call electron orbitals, such as the $1s$ and $2p_z$.

As you might expect since we have a system with an electrostatic potential energy function, the energy of the electron in the atom depends only on the separation distance, r, of the electron from the positive nucleus. We can, using the principles

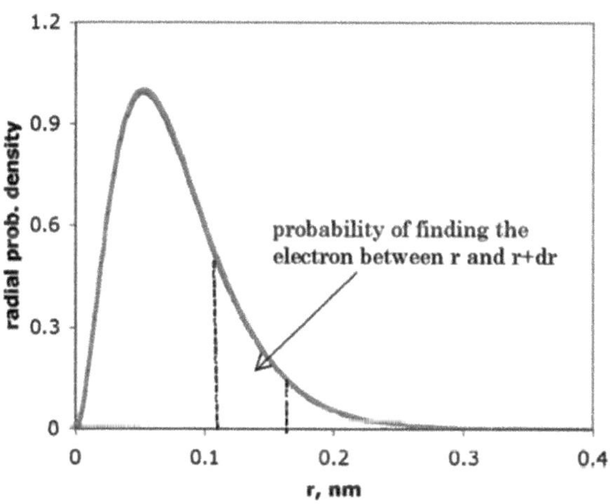

Figure 6.6. Use of the radial probability distribution function to calculate probability.

of calculus, calculate the average position of the electron in the $1s$ orbital from the expression:

$$g_{\text{avg}} = \int_0^{\infty} PgPdg = \int_0^{\infty} P^2 gdg$$

The integrals for the electron in hydrogen are more complicated than we really wish to address here, so let's use the probability density for the one-electron particle in a box with $V = 0$, when the particle is in the box:

$$\Psi_{n=1}(x) = \left(\frac{2}{L}\right)^{1/2} sin\left(\frac{\pi x}{L}\right).$$

Recall that the box extends from $x = 0$ to $x = L$, so to find x_{avg}:

$$\begin{aligned} x_{\text{avg}} &= \int_0^L \left(\frac{2}{L}\right)^{1/2} sin\left(\frac{\pi x}{L}\right) x \left(\frac{2}{L}\right)^{1/2} sin\left(\frac{\pi x}{L}\right) dx \\ &= \frac{2}{L}\int_0^L sin^2\left(\frac{\pi x}{L}\right) x\, dx \\ &= \frac{2}{L}\cdot\frac{L^2}{4} = \frac{L}{2} \end{aligned}$$

So the average position of the particle in the box, when $n = 1$, is in the center of the box or at $L/2$.

What is the most probable position of the electron? Again, going back to calculus, we can begin to find the most probable position by taking the first derivative of the probability density, then finding x_{mp} by setting the derivative equal to zero and solving for x_{mp}:

$$\frac{d}{dx}\Psi^2_{n=1}(x) = \frac{d}{dx}\left(\frac{2}{L} sin^2\left(\frac{\pi x}{L}\right)\right) = 0$$

$$x_{mp} = \frac{L}{2}$$

Let's find r_{mp} for the 1s electron in the hydrogen atom. The total wavefunction is:

$$\Psi_{1s} = \left(\frac{1}{4\pi}\right)^{1/2} * 2\left(\frac{1}{a}\right)^{1/2} e^{-r/a} \quad \text{and} \quad r^2\, R^2(r) \quad \text{is} \quad \frac{4}{a} r^2 e^{-2r/a}$$

So,

$$\frac{d}{dr} r^2 R^2(r) = \frac{-2}{a} r^2 e^{-2r/a} + 2re^{-2r/a} = 0$$

$$\frac{-2}{a} r^2 + 2r = 0 \rightarrow r_{mp} = a, \quad \text{where } a = a_0, \text{ the Bohr radius}$$

We can also calculate the position of nodes using the wavefunctions. For $n = 2$, the particle-in-a-box wavefunction is:

$$\Psi_{n=2}(x) = \left(\frac{2}{L}\right)^{1/2} sin\left(\frac{2\pi x}{L}\right)$$

Because a node means that both the wavefunction and the probability density both equal zero, we can work with the wavefunction and find L such that:

$$\Psi_{n=2}(x) = \left(\frac{2}{L}\right)^{1/2} sin\left(\frac{2\pi x}{L}\right) = 0$$

$$sin\left(\frac{2\pi x}{L}\right) = 0 \quad \text{when} \quad \left(\frac{2\pi x}{L}\right) = \pi \quad \text{or}$$

$$x = \frac{L}{2}, \text{ as seen in Figure 5.2}$$

Another example is to find the node in the 2s one-electron radial wavefunction:

$$\text{R}_{2s} = \left(\frac{1}{a_0}\right)^{3/2} \left(2 - \frac{r}{a_0}\right) e^{-r/2a_0}$$

This function will equal zero when $(2 - \frac{r}{a_0}) = 0$ or $r = 2a_0$.

A complete list of the radial and angular hydrogen ($Z = 1$) wavefunctions for $n - 1$, 2, and 3 follows. These will be useful in considering the shapes of one-electron orbitals in Chapter 7.

Angular Wavefunctions, $Y(\theta, \phi)$

s orbitals ($l = 0$)

$$Y_{l=0} = \left(\frac{1}{4\pi}\right)^{1/2}$$

p orbitals ($l = 1$)

$$Y_{l=1} = \left(\frac{3}{4\pi}\right)^{1/2} sin(\theta)cos(\phi);\ p_x$$

$$Y_{l=1} = \left(\frac{3}{4\pi}\right)^{1/2} sin(\theta)sin(\phi);\ p_y$$

$$Y_{l=1} = \left(\frac{3}{4\pi}\right)^{1/2} cos(\theta);\ p_z$$

d orbitals ($l = 2$)

$$Y_{l=2} = \left(\frac{5}{16\pi}\right)^{1/2} \left(3cos^2(\theta) - 1\right);\ d_{z^2}$$

$$Y_{l=2} = \left(\frac{15}{4\pi}\right)^{1/2} sin(\theta)cos(\theta)cos(\phi);\ d_{xz}$$

$$Y_{l=2} = \left(\frac{15}{4\pi}\right)^{1/2} sin(\theta)cos(\theta)sin(\phi);\ d_{yz}$$

$$Y_{l=2} = \left(\frac{15}{16\pi}\right)^{1/2} sin^2(\theta)sin(2\phi);\ d_{xy}$$

$$Y_{l=2} = \left(\frac{15}{16\pi}\right)^{1/2} sin^2(\theta)cos(2\phi);\ d_{x^2-y^2}$$

Radial Wavefunctions, $R(r)$

$1s$ orbital $R(r) = 2\left(\frac{1}{a_0}\right)^{3/2} e^{-r/a_0}$

$$2s \text{ orbital } R(r) = \frac{1}{2\sqrt{2}}\left(\frac{1}{a_0}\right)^{3/2}\left(2 - \frac{r}{a_0}\right)e^{-r/2a_0}$$

$$3s \text{ orbital } R(r) = \frac{2}{81\sqrt{3}}\left(\frac{1}{a_0}\right)^{3/2}\left(27 - 18\frac{r}{a_0} - 2\left(\frac{r}{a_0}\right)^2\right)e^{-r/3a_0}$$

$$2p \text{ orbital } R(r) = \frac{1}{2\sqrt{6}}\left(\frac{1}{a_0}\right)^{3/2}\left(\frac{r}{a_0}\right)e^{-r/2a_0}$$

$$3p \text{ orbital } R(r) = \frac{4}{81\sqrt{6}}\left(\frac{1}{a_0}\right)^{3/2}\left(6\left(\frac{r}{a_0}\right) - \left(\frac{r}{a_0}\right)^2\right)e^{-r/3a_0}$$

$$3d \text{ orbital } R(r) = \frac{4}{81\sqrt{30}}\left(\frac{1}{a_0}\right)^{3/2}\left(\frac{r}{a_0}\right)^2 e^{-r/3a_0}$$

Exercises

1. What are the quantum numbers for the second excited state of the hydrogen atom?
2. If the radial probability density plots in Figure 6.3 were not labeled, how would you determine which curve represented the $2s$ orbital?
3. Show that the definition of the Cartesian coordinates in spherical polar coordinate form (Equation (6.2)) results in $r = \sqrt{x^2 + y^2 + z^2}$.
4. Why is the average value of r (Equation (6.6)) a useful quantity?
5. Are the energies of the eight sets of quantum numbers for the $n = 2$ levels of the hydrogen different? Why, or why not?
6. What is the most probable position of the electron in the hydrogen atom $1s$ orbital?
7. Describe how you would locate the node in the hydrogen atom $2s$ orbital.
8. Does the $2p_z$ angular wavefunction have a node? Does the $2p_z$ radial wavefunction?

CHAPTER 7

Shapes of Orbitals

Those pictures of orbitals you see in every chemistry or physics book are not shapes that have been measured or observed in some way. There is no scientific instrument that can directly measure the shape of an electron orbital. The best we can do is to generate a "signature" of an orbital using very sophisticated instrumentation. The reason is Heisenberg's Uncertainty Principle.

Recall that we cannot image an object unless the wavelength of radiation we are using is just smaller than the object itself. Of course, that implies that we will transfer meaningful momentum to the object and it will change position. We must treat the electron as a wave, not a particle. To draw an analogy with the particle in a box, we are trying to find out where the particle is by shaking the box.

So from where do all these pictures of s, p and d orbitals arise? They are three-dimensional graphs of the solutions to Schrödinger's equation corresponding to different choices of eigenvalues and different solutions corresponding to each choice of eigenvalue.

A solution to Schrödinger's equation is a function of three spatial variables that describe the probability distribution of the position of an electron as a wave form. We picture it as a cloud in three-dimensional space. The darker the cloud, the higher the distribution function is in that region and the more likely the electron is located there. That is why the orbitals are always portrayed as clouds.

Now, what do we learn from the time-independent solution to Schrödinger's equation? We learn that we can construct solutions of the form:

$$\Psi(x, y, z) = R(r)Y(\theta, \phi) \tag{7.1}$$

In such cases $Y(\theta, \phi)$ is a spherical harmonic (but could be further separated into $\Theta(\theta)$ and $\Phi(\phi)$) and R is a product of a power of r, an exponential function, and a Legendre polynomial in r. We want to take a snapshot, so time is fixed and we only care about the spatial coordinates. If we want to visualize these solutions it is useful to think about where the functions are zero and what sorts of symmetry are inherent in the solutions.

We begin with the simplest case, the energy level corresponding to $n = 1$. You should recall that $l = 0$ for the $n = 1$ principal quantum number. The Legendre polynomial is zero so this leaves only the decreasing exponential in r in $R(r)$ to examine. That is, the *entire* solution depends only on r and decreases exponentially away from the origin (where we placed the nucleus of the atom). This is how we get our standard picture of the $1s$ orbital as shown in Figure 7.1.

It is necessary to point out a slight annoyance of notation here. Physicists and chemists call the principal quantum number n but mathematicians use n for the degree of the polynomial that is contained in $Y(\theta, \phi)$. We will use n for the principal quantum number, but be mindful that the mathematical supplemental

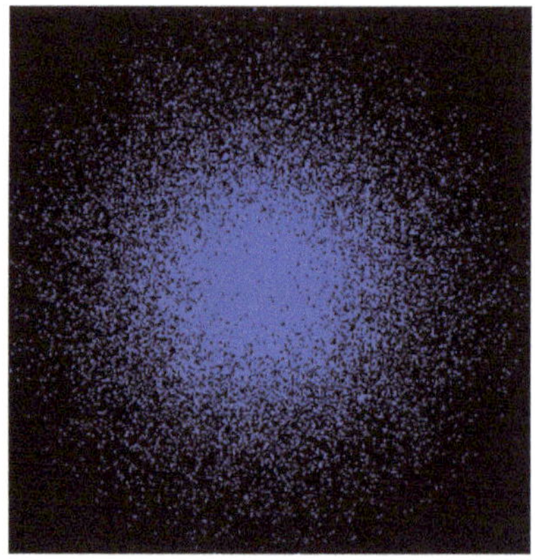

Figure 7.1. The $1s$ orbital; the nucleus is at the center of the plot.

chapters use a different notation. You can also see from this analysis that this is the only solution corresponding to $n = 1$. That is, there is only one possible orbital shape for the electron in the lowest energy level of hydrogen.

All of the s orbitals correspond to solutions where $Y(\theta, \phi)$ is constant. So the next one we want to look at is the $2s$ orbital. The principal quantum number n is equal to 2. Our solution is still given only in terms of r, but now the associated Legendre polynomial has degree 1 and there is one value of r for which it is zero, i.e., there is a node in this orbital. This gives a picture with concentric shells of positive density for the $2s$ orbital (Figure 7.2).

A similar discussion leads to pictures of the $3s$, $4s$, and $5s$ orbitals, all with more and more shells of positive density as the degree of the Legendre polynomial associated with $R(r)$ goes up as shown in Figure 7.3.

Of course, if the principal quantum number is 2 then it is also possible to define an electron density cloud by setting $n = 2$ and $l = 1$. When $l = 1$, the function $R(r)$ is still decreasing exponentially. But if the degree of the polynomial that gives us $Y(\theta, \varphi)$ is 1, then $Y(\theta, \varphi)$ might come from any combination

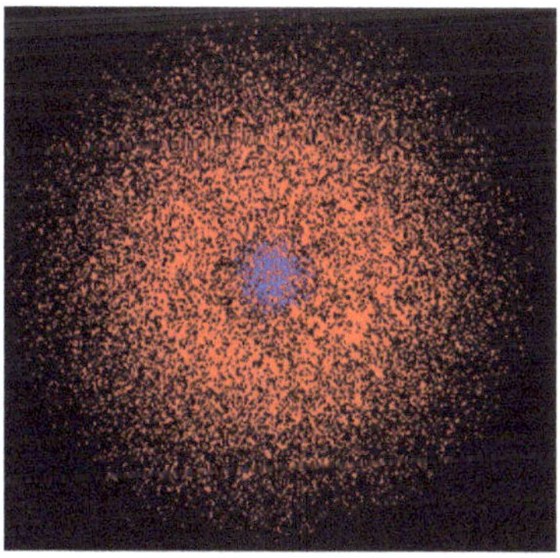

Figure 7.2. The $2s$ orbital. The transition from red to blue is a node.

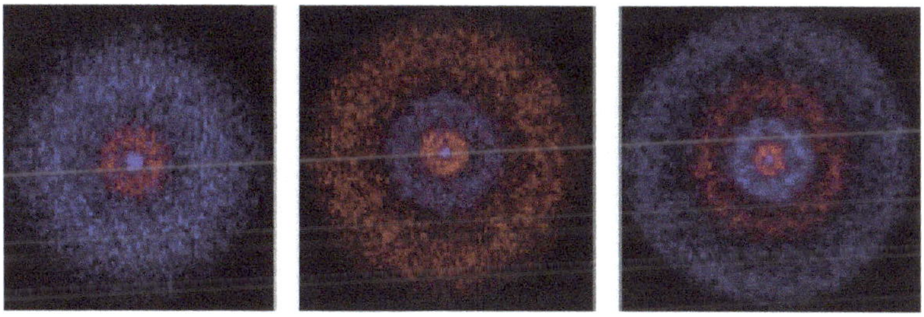

Figure 7.3. The $3s$, $4s$, and $5s$ orbitals. Note the increase in the number of nodes as the principal quantum number n increases.

of three possibilities: x, y or z. The solutions corresponding to these three are given by the relation:

$$\begin{aligned} x &= r\, sin(\theta)cos(\phi) \\ y &= r\, sin(\theta)sin(\phi) \\ z &= r\, cos(\phi) \end{aligned} \tag{7.2}$$

so setting $r = 1$ yields three distinct candidates for $Y(\theta, \phi)$:

$$\begin{aligned} Y(\theta, \phi) &= sin(\theta)cos(\phi) \\ Y(\theta, \phi) &= sin(\theta)sin(\phi) \\ Y(\theta, \phi) &= cos(\theta) \end{aligned} \tag{7.3}$$

Figure 7.4. The $2p_z$ orbital. The nucleus is located at the node.

If we look at the last of these, $Y(\theta, \phi) = cos(\theta)$, we can ask where this expression is zero and where it is largest. We know that $cos(\theta) = 0$ when $\theta = \pi/2$ or 90° down from the positive z axis. That is, $Y(\theta, \phi)$ is zero exactly when z is zero, or on the xy plane. If you take this piece of information and put it with the exponentially decreasing $R(r)$, you get a two-lobed orbital lined up with the z axis as shown in Figure 7.4.

This is the $2p_z$ orbital. There are three $2p$ orbitals considered possible, and they are given by the three solutions distinguished above. The solution that comes from the polynomial x, $Y(\theta, \phi) = sin(\theta)cos(\phi)$, is zero when x is zero. You can use the properties of the polynomial to get this or you can actually work it out from the angular variables. Figure 7.5 just looks like a rotated version of the last solution in Figure 7.4.

Because of Schur's lemma, we should not be surprised by this phenomenon. Any rotated version of a solution to Schrödinger's equation will yield another solution with the same eigenvalue (or energy level, or principal quantum number). So, of course, there is a third $2p$ orbital as shown in Figure 7.6.

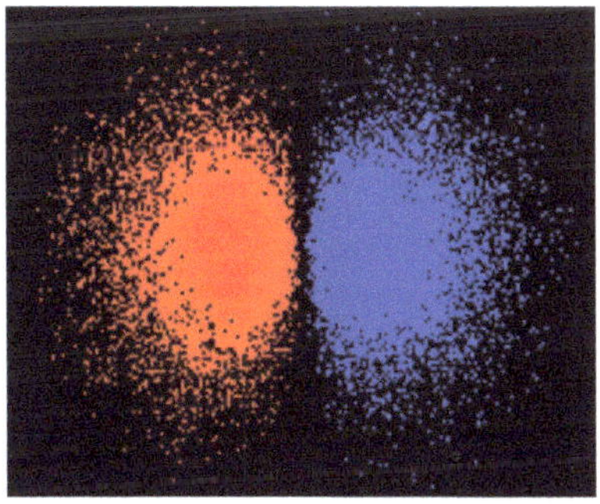

Figure 7.5. The $2p_x$ orbital.

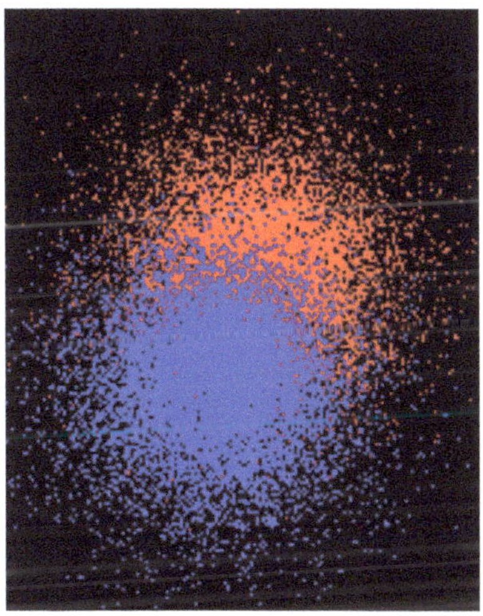

Figure 7.6. The $2p_y$ orbital.

This one is aligned with the y axis. But any orientation of a $2p$ orbital gives some kind of $2p$ orbital, so the choice of these three as standard candidates is somewhat arbitrary. On the other hand, the number 3 is the dimension of the vector space of solutions corresponding to $l = 1$, so it is a very important number. Chemists use this number to determine the number of electrons that can fit in a given atomic "shell". The number of orbitals in the shell corresponding to the principal quantum number 2 is four — one $2s$ orbital and three $2p$ orbitals.

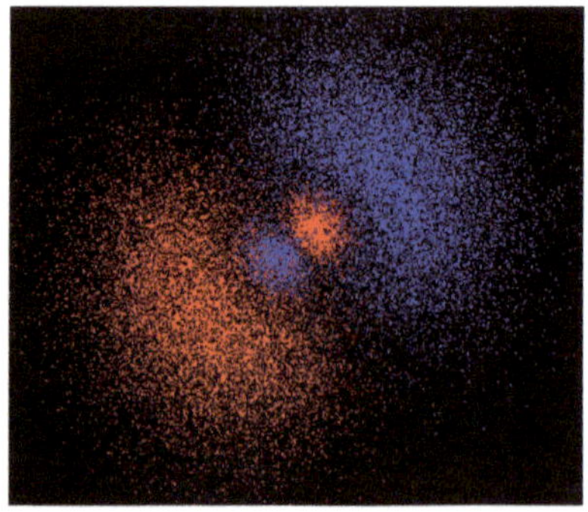

Figure 7.7. The $3p_y$ orbital. Unlike the $2p_y$ orbital, this orbital has a node.

What happens if the principal quantum number is 3? There are three ways to achieve this, with $l = 2$, 1, or 0. If $l = 0$ you get the $3s$ orbital as in Figure 7.3. If $l = 1$ you get the same solutions for $Y(\theta, \phi)$ as you would in the case of the $2p$ orbital. Now, however, these are multiplied by Legendre polynomials of degree 1. These introduce a node at a fixed value of r. This cuts the p orbital in two by creating a sphere of zero density at this particular value of r. The result is the $3p$ orbitals, of which there are 3 linearly independent choices (the rest being made up by linear combinations of these three). A typical one looks like Figure 7.7. The rest are all rotated versions of this one. Finally we should consider the case where $n = 3$ and $l = 2$. Now there will be no radial nodes because the Legendre polynomial is constant. All the action will be in the angular variables. When $l = 2$ there are several harmonic polynomials that are easy to check: xy, yz, xz. It is easy to see where these are zero. The first one, for example, is zero if $x = 0$ or if $y = 0$. These are the yz and xz planes, respectively. So the solution corresponding to xy has 4 lobes in each of the 4 quadrants of the xy plane, The function $R(r)$ ensures that the distribution dies off exponentially away from the origin. The result gives a $3d$ orbital (Figure 7.8).

One can rotate this picture to give the other two easy harmonic polynomials of degree 2, yz and xz. But this is not the

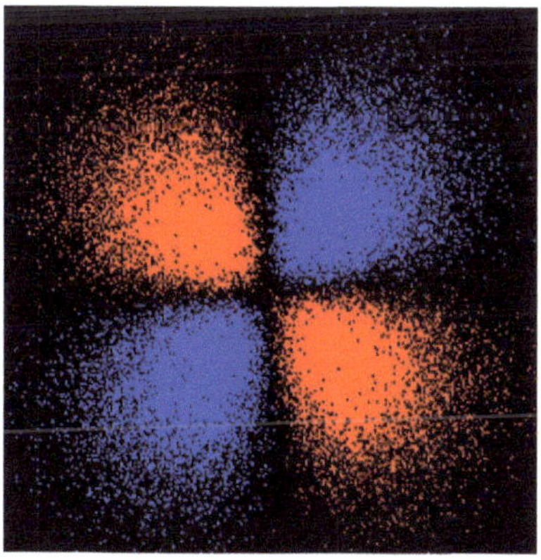

Figure 7.8. The $3d_{xy}$ orbital. This orbital, the lowest of the d orbitals, has two nodes. Recall that the lowest s orbital, the $1s$, has zero nodes and the lowest p orbital, $2p$, has one node.

whole story because the dimension of the space of homogeneous polynomials of degree two is 6 (given by combinations of xy, xz, yz, xx, yy, zz) and we lose one dimension by setting:

$$x^2 + y^2 + z^2 = 1 \tag{7.4}$$

So we would expect five distinguishable solutions altogether. But recall that there are some subtle ones such as:

$$z^2 - \frac{1}{3}(x^2 + y^2 + z^2) \tag{7.5}$$

which reduces to $z^2 - \frac{1}{3}$ (7.6) on the unit sphere. When this is put in spherical coordinates it is given by:

$$G(\theta, \phi) = cos^2(\theta) - \frac{1}{3} \tag{7.6}$$

This solution is zero exactly when θ is at some fixed angle. So the node is a cone perpendicular to the z axis at that fixed angle. The solution corresponding to this polynomial has lobes along the z axis and also a donut-shaped cloud around the z axis below the node as shown in Figure 7.9.

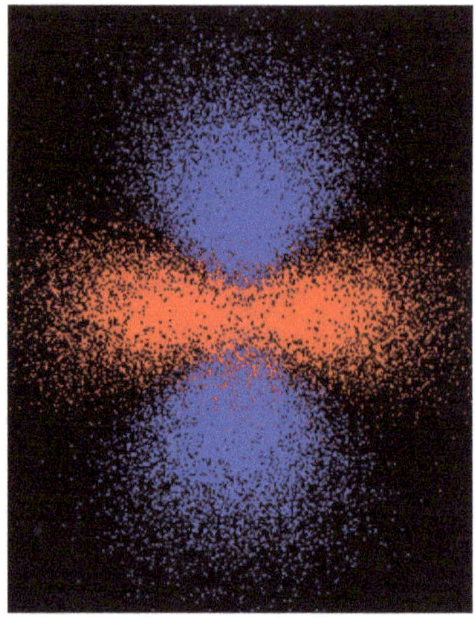

Figure 7.9. The $3d_{z^2}$ orbital.

You can see from the formula that the solution doesn't depend on ϕ, therefore you see a circular symmetry around the z axis. This is one of the d orbitals frequently pictured in texts. There are two linearly independent forms of this for a total of five possible d orbitals, no matter what energy level. So for principal quantum number 3, there are 5 d orbitals, 3 p orbitals and 1 s orbital for a total of 9 possible states. Of course you can continue this analysis for as long as you want, generating pictures and formulas for orbitals of arbitrary quantum number and complexity. Although in what follows we will generally not use more than the s and p orbitals, it is good to know that the mathematical machine that produced them is capable of producing many, many more.

Before we proceed with our discussion of molecules it might be useful to point out how typical this development is in science. We have replaced our axiomatic model of electrons in shells with a much more sophisticated model derived from Schrödinger's equation and its specific solutions. This allows us to use the shapes of orbitals, which are really the shapes of waveforms solving a particular equation, to predict bonding angles and other properties. But we can't see the orbitals directly by any scientific experiment. They are a mathematical construction, not an observable fact. Yet they are close enough to being correct to allow us to improve our model of how atoms bond. In fact, they represent a huge breakthrough in how we look at these very small things. What the mathematics has done is to allow our main assumption (the electron is a wave) its full predictive power. These atomic orbital shapes are crucial in the development of an analogous system for molecules. Not surprisingly, that systematic study of molecules is known as Molecular Orbital-Linear Combination of Atomic Orbital Theory, or MO (Molecular Orbital) theory for short. We will explore that model in Chapter 10.

CHAPTER 8

Molecules and Simple Bonding Models

(Lewis Dot Structures)

We noted that multi-electron atoms were complicated and that we could not analytically solve the Schrödinger equation for that case without using approximations. Treating the electrons in molecules is, obviously, even more complicated. However, as chemists we can first develop simple models to explain some aspects of bonding before using quantum chemistry to treat such a system as best we can with our approximations. In these next chapters we will examine useful methods to understand some aspects of chemical bonding.

The number of known chemical compounds, molecules, is staggering. Clearly, there is a propensity for atoms to combine into larger units. Can we apply the quantum ideas we previously developed to this problem? As you already know, an instructor never asks this type of rhetorical question unless the answer is yes. So one of the basic questions we must address, regardless of our bonding theory, must be: "Why does bonding occur?" We might expect to find the answer in thermodynamics, but as noted above let's put off explicitly answering that question until we make a few observations:

(1) The H–O–H bond angle is approximately the same (~109°) regardless of the molecule in which it occurs, be it water or methanol.
(2) Bond energy, for a given type of bond, does not appear to differ greatly from molecule to molecule. The energy of the C–H bond in CH_4 is ~103 kcal mole^{-1}, while in C_2H_6 it is 96 kcal mole^{-1}.

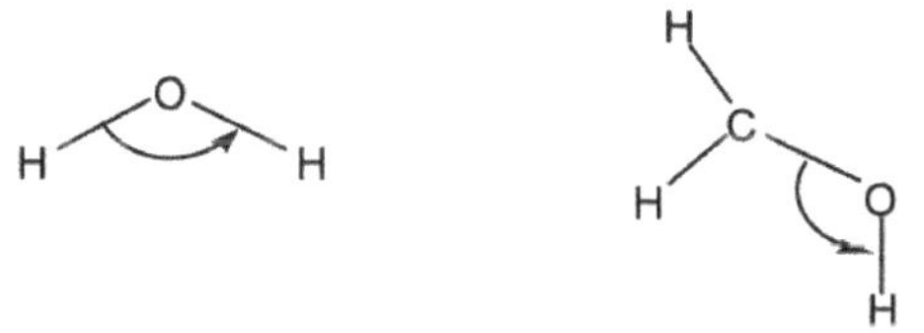

(3) Elements in a given column of the periodic table have similar bonding characteristics. C, Si and Ge all form four bonds with H and all have H–X–H bond angles of ~109°.

It seems as if these brief observations have provided us with additional criteria for any bonding theory that we might consider. That is, a successful theory of bonding must account for the molecular formula and the molecular geometry, as well as the driving force for bonding.

To complicate matters even more, we note that while we all have some familiarity with the terms covalent bond and ionic bond, these are not the exclusive set of bonding types. In fact, few molecules are purely covalent or purely ionic. In addition to these chemical bonds, there are non-bonding interactions among atoms to provide what are generically known as weakly bound species. Included among this group are molecules with hydrogen bonding. The remainder of this monograph will describe bonding models and theories, then address hydrogen bonding and, finally, other weak interactions.

Lewis Dot Structures are the simplest approach to rationalizing chemical bonds. There are three different versions of this elementary idea. We start with the original as stated by Lewis himself in 1916:

> "...atoms strive to achieve inert gas electron distributions by sharing electrons."

Note that no distinctions are made between s and p electrons, because no one knew about such things in 1916. For light atoms, the model was based on the number of electrons as determined by the atomic numbers of the atoms. We represent atoms by dot structures, with one dot per electron, and we restrict ourselves to first-row atoms. I left as many unpaired electrons as I could.

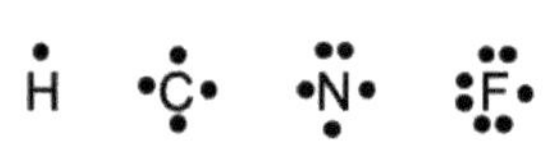

We represent molecules by bringing such atoms together to make bonds, where a dash represents two electrons in a bond. In these molecules, each atom has an inert gas configuration; for H in H_2 it is that of He and for F in HF, it is that of Ne.

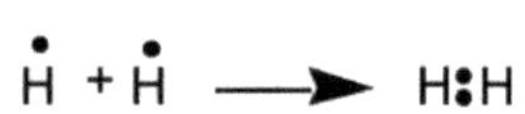

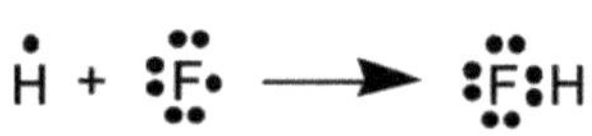

The second form of Lewis's model is known as the "octet rule" and is stated as:

> "...except for H (and Li, Be and B), atoms form bonds until surrounded by eight electrons."

This "rule" is only valid for atoms with $Z \leq 10$ and shows why ammonia has 3 N–H bonds. We know about orbitals, so what happens as $Z > 10$? We move from the $n = 1$ and $n = 2$

orbitals containing only s and p electrons to principal quantum number $n = 3$, where d electrons are possible.

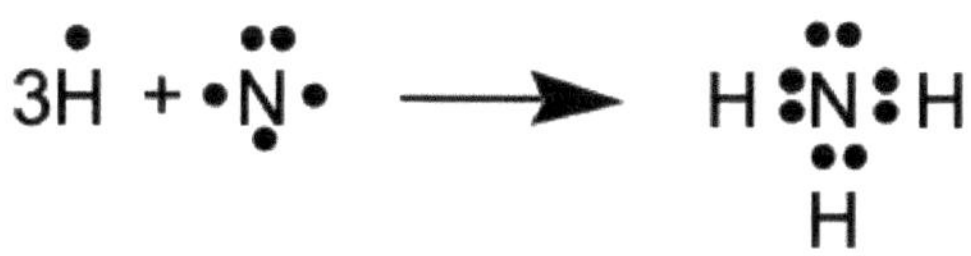

Look at the reaction between P and Cl. Phosphorous forms a molecule with three P–Cl bonds. However, it also forms a second molecule:

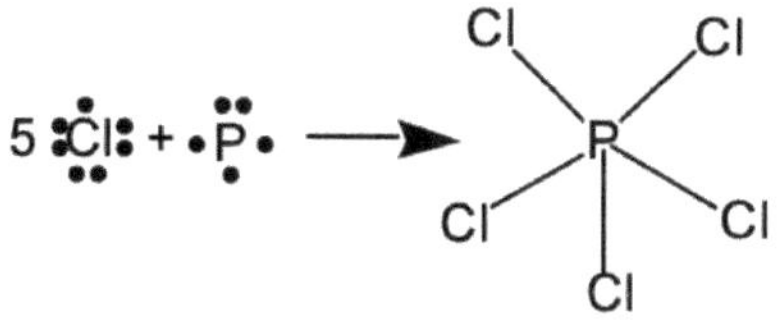

The phosphorous atom in this molecule has five bonds and 10 electrons associated with it, in clear violation of the octet rule. However, we must remember that P has $Z = 15$, so there is no reason to believe that the octet rule will be obeyed. Sulfur has one more electron than phosphorous and we expect that it could form six bonds in a Lewis dot structure. This leads to the last (and best) version of the Lewis dot structure rules which acknowledges the existence of different orbitals:

> "...form as many bonds, at least there is the potential to form as many bonds, as there are available electrons."

We define available electrons as the maximum number of unpaired electrons. This is not the standard definition! Going back to the PCl_5 example, the phosphorous atom may have 5 unpaired electrons because d orbitals are available. For the related N atom, no d orbitals are available ($n = 2$), so the maximum number of unpaired electrons is 3. We can also return to the carbon example cited at the beginning of this section. The four outer shell electrons can be distributed among the four s and p total orbitals, so carbon can form four bonds.

The best way to see what all of the above text implies is to try it out. We'll do that by looking at some real molecules. Our first example is the freon molecule tetrafluoromethane, CF_4. The total number of available electrons (32) is given by the carbon contribution of 4 electrons and the contribution of 7 electrons from each fluorine. The structure we draw follows the octet rule.

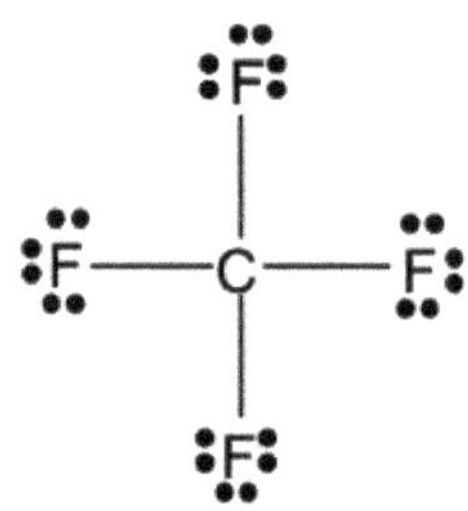

Molecular nitrogen, N_2, is a slightly more complicated example. Each nitrogen atom contributes 5 electrons, so the total number of available electrons is 10. Begin by drawing a structure that obeys the octet rule while ignoring the total number of electrons.

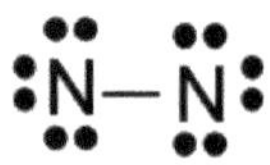

This requires 14 electrons which is four electrons too many. Therefore, we replace the single bond with a double bond. This reduces total number of electrons in our dot structure by two.

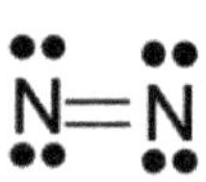

We still have too many, 12, electrons in the structure, so we add another bond.

This works! The lesson to take from this example: adding a double bond reduces the total number of electrons by 2 and a triple bond reduces that number by 4. The steps, shown in N_2, exemplify the general procedure to follow in writing Lewis dot structures.

The final example will require us to pursue a new concept. The molecule is the nitroamide anion, $N_2HO_2^-$. The number of available electrons, 24, is given by the sum of those from N (10), H (1), O (12) and the charge (1). Our trial structure would be

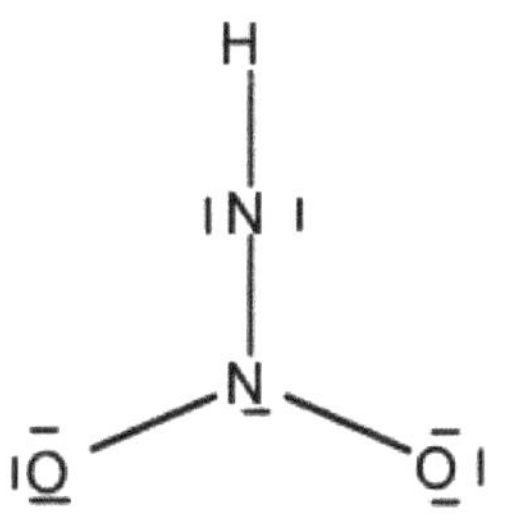

which requires 26 electrons. We need to add one double bond to our trial structure, but there are three possible positions for the double bond, as shown in the structures below. These structures are known as **resonance structures**. The true structure is some weighted average of these three. That is, each N–O bond is between a single and double bond in length, as is the N–N bond. In order to rank the relative importance of each structure, we need to use **formal charges**.

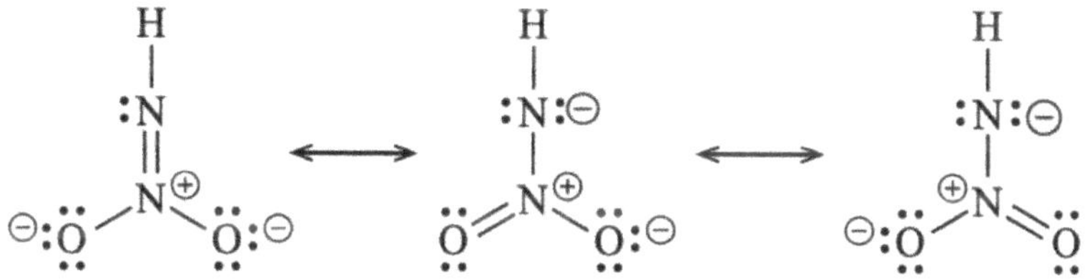

(I) (II) (III)

Formal charge, FC, is an arbitrary property. We define the formal charge on a specific atom by:

- number of outer shell electrons on the free atom
- − number of non-bonded electrons on the atom in the molecule

- $-1/2$ of the number of bonding electrons around the atom.

For structure (I),

$$\mathrm{FC_H} = 1 - 0 - \frac{1}{2}(2) = 0$$

$$\mathrm{FC_O} = 6 - 6 - \frac{1}{2}(2) = -1$$

$$\mathrm{FC_N} = 5 - 2 - \frac{1}{2}(6) = 0$$

$$\mathrm{FC_N} = 5 - 0 - \frac{1}{2}(8) = 1$$

For structures (II and III),

$$\mathrm{FC_H} = 1 - 0 - \frac{1}{2}(2) = 0$$

$$\mathrm{FC_N} = 5 - 4 - \frac{1}{2}(4) = -1$$

$$\mathrm{FC_N} = 5 - 0 - \frac{1}{2}(8) = 1$$

$$\mathrm{FC_O} = 6 - 6 - \frac{1}{2}(2) = -1$$

$$\mathrm{FC_O} = 6 - 4 - \frac{1}{2}(4) = 0$$

We have three guidelines to assist in the assignment of the contributions of each structure to the observed structure:

(i) Electroneutrality is preferred, so that the "best" structure has no formal charge at all.
(ii) We wish to minimize the formal charges, so that charges of ± 1 are preferable to ± 2, etc. Negative charges are preferably positioned on more electronegative elements.
(iii) Opposite charges should be as close as possible, while like charges should be as far apart as possible.

Based on these criteria, we conclude that structure (I) is more important or contributes a larger fraction to the observed structure. We base this on assigning negative formal charges to oxygen rather than nitrogen atoms. The final assignment is that (I) > (II) = (III).

We conclude by asking 'how well does the Lewis dot structure formalism satisfy our criteria for a satisfactory bonding theory?' The obvious answer is: not very well. It will predict the correct molecular formula, but does not address geometry or the rationale for bonding.

Exercises

1. Write Lewis dot structures for the atoms and ions: Ne, Na^+, B and O.
2. Write Lewis dot structures for the isoelectronic (having the same number of electrons) species: N_2, CN^- and CO.
3. Write Lewis dot structures for the molecules: SiH_4, SO_2, NO^+, OCl^-, COS and NO_2^-.
4. Write a Lewis dot structure for the SF_6 molecule.
5. Write Lewis dot structures for the following molecules/ions: N_2O_4, C_2H_4 and SO_4^{2-}.

CHAPTER 9

Molecules and Simple Bonding Models II

(Valence Shell Electron Pair Repulsion)

We have used Lewis dot structures to provide, at least in some cases, the formula for a molecule. We can begin with that information and develop a model for molecular geometry based on the molecular formula. One way to deal with the specific geometry of a molecule with a known formula is to return to Coulomb's law, that is, to look at electron-electron repulsion. Valence Shell Electron Pair Repulsion, VSEPR, is such an electrostatic theory of bonding. As with Lewis dot structures, it ignores specific orbitals, but in a very defined manner. The observed geometry is expected to reflect the intrinsic attempt to minimize electron-electron repulsion by maximizing the distance between electrons. Electrons comprising bonds are highly localized in the space between the bonded atoms. The VSEPR model assumes that bond angles are determined solely by the number of valence electrons around a central atom. Valence electrons are defined, in this case, as the number of electrons in the outermost shell provided by the principal quantum number. It is instructive to use examples.

(1) BeH_2: Be has four total electrons, but only two electrons in the outermost or $n = 2$ shell. For model purposes we note that Be has two valence electrons, so that the maximum separation between two electrons when the hydride is formed is 180°, leading to a linear geometry.

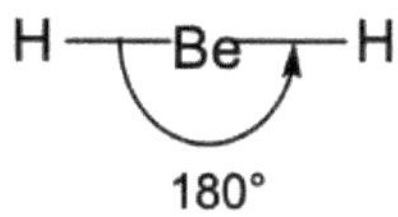

(2) BH_3: B ($1s^22s^22p$) has three valence or $n = 2$ electrons with a maximum separation of 120°, leading to a trigonal molecular geometry in the hydride.

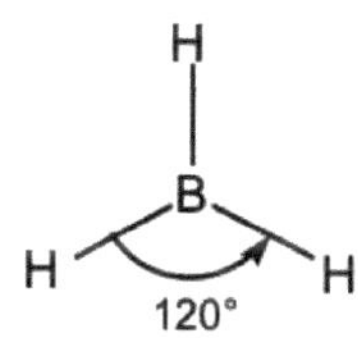

(3) SF_6: S has an "expanded" octet, that is, it can "use" d orbitals to "unpair" electrons, at least in our thought experiment, since $Z > 10$. S ($1s^22s^22p^63s^23p^43d^0$) has six available or valence electrons and can form six bonds with fluorine, 90° apart, with octahedral geometry.

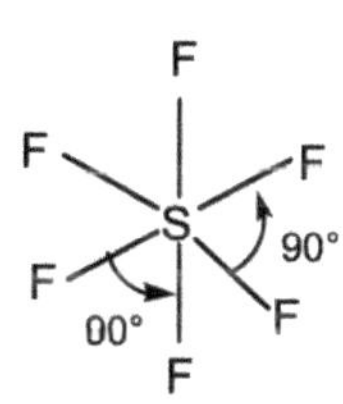

(4) We saw PCl_5 earlier. The bond angles for maximal electron separation are not equivalent. The PCl_5 geometry is called trigonal bipyramidal with three 120° angles and two 90° angles.

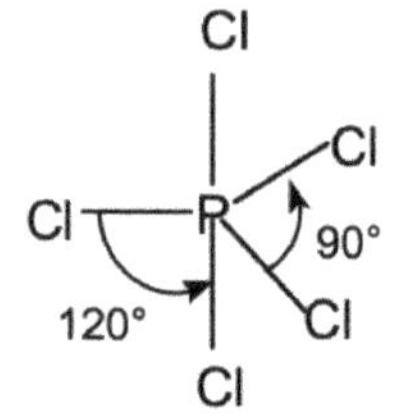

Recall an earlier Lewis dot structure discussion of the bonding in ammonia. The nitrogen atom ($1s^22s^22p^3$) has a non-bonding pair of electrons that cannot be separated even in a thought experiment. How does this change the geometry according to the VSEPR model? We treat the pair of electrons as a bond for purposes of assigning an ideal or hypothetical geometry. We consider NH_3 to have four "bonds". The maximum separation would arrange the bonds in a regular tetrahedron with bond angles of 109.5° just as in CH_4. However, electrons not involved in bonding are not as localized in space and distort the ideal geometry. The ammonia H–N–H bond angle is less than 109.5°, approximately 107°. The observed ammonia geometry is pyramidal. Similar geometric distortions are observed in the experimental geometries of H_2O and SF_4.

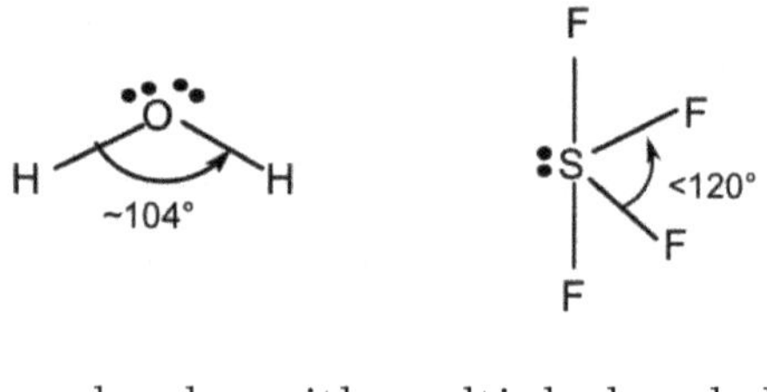

We also saw molecules with multiply bonded atoms, double and triple bonds. What effect do multiple bonds have on the VSEPR geometry? For purposes of geometry considerations, we simply treat double and triple bonds as single bonds. Consider ethylene, C_2H_4.

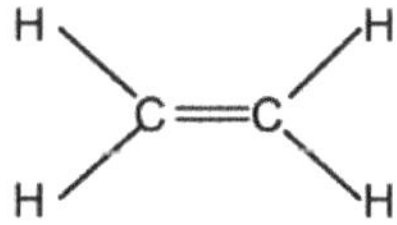

The geometry around each carbon is trigonal planar, as though only three bonds were present.

Let's summarize VSEPR: For molecules with a central atom, the bonding electrons arrange in a geometry that minimizes electron-electron repulsion. In these molecules, pairs of non-bonding electrons experience the greatest repulsion because of their relative freedom of movement. Similarly, the interaction of non-bonding electrons with bonding electrons is greater than the interaction between two pairs of bonding electrons. We can predict the geometry from the number of bonds as described in the examples above. For example, BeH_2 is linear and CH_4 is tetrahedral. PCl_5 and SF_6 present geometries we may not have encountered previously, trigonal bipyramidal and octahedral, respectively. In considering possible molecular geometries, we treat non-bonding electrons as a "bond" (NH_3, for example, has a pyramidal geometry based on a perfect tetrahedral structure). We also treat the presence of multiple bonds as a single bond (C_2H_2, for example, is linear).

It is time to reflect on how well VSEPR satisfies the criteria for a "successful" bonding theory. Since it combines the features of Lewis dot structures (molecular formula) with a geometric prediction, we fulfill two of the three requirements, but we still have not answered the major question, that is, the rationale for bonding. Our search for an explanation of the tendency of atoms to combine into molecules must continue.

Exercises

1. Write Lewis dot structures and predict the geometry of the molecule (the central atom is in bold type): $H_2\mathbf{O}$, $\mathbf{Xe}F_4$, $\mathbf{S}F_6$, $\mathbf{Cl}F_3$.

2. Write Lewis dot structures and predict the geometries of the following molecules and ions: $\mathbf{CO_2}$, $\mathbf{SO_2}$, $\mathbf{CS_2}$, H_2CO.
3. Write Lewis dot structures and indicate the geometry of each of the following: $\mathbf{AsF_3}$, **N**NO.
4. The nitrosyl fluoride molecule, NOF, is used as a rocket propellant. In principle, either O or N could be the central atom. However, nitrogen has a lower ionization potential, so that should be the central atom. Draw a Lewis structure for this molecule and describe its geometry.

CHAPTER 10

Molecular Orbital Theory

Our quest leads now to a model that is quantitative: Molecular Orbital Theory (MO Theory). This is an extension of the atomic orbital theory we have recently encountered. If it is acceptable for atoms, we should be able to extend it or an analogous model to molecules. Consider the simplest molecule, H_2^+, a one-electron system, the molecular analogy to the hydrogen atom. Begin with a thought experiment. Imagine that we could measure the probability density Ψ^2, that is, the probability per unit volume of finding the electron in H_2^+. If we construct a plot of Ψ^2 as a function of the distance from the hydrogen atoms, we would see the plot shown in Figure 10.1.

The probability of locating the electron on a line between the two nuclei is substantial. $\Psi^2_{H_2^+}$ is symmetric about both atoms. It is the molecular analog of the hydrogen atomic orbital, it is a molecular orbital (MO) that we designate as a "sigma" orbital, σ. Now compare $\Psi^2_{H_2^+}$ with $\Psi^2_H + \Psi^2_H$ (the unbonded case) in Figure 10.2. Formation of the bond increases the probability

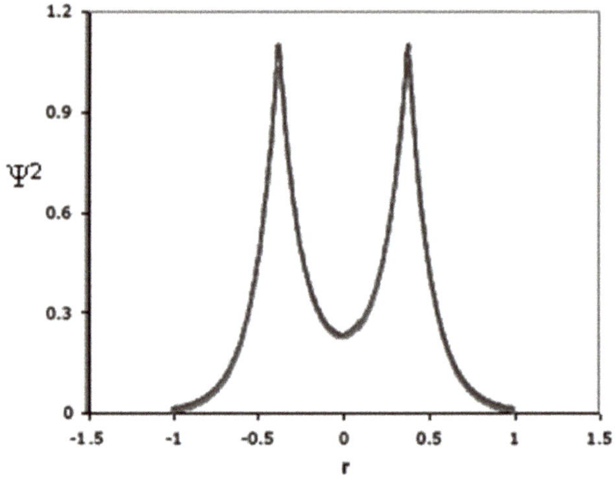

Figure 10.1. σ_{1s} orbital for H_2^+. The hydrogen nuclei are located at the peaks in the distribution.

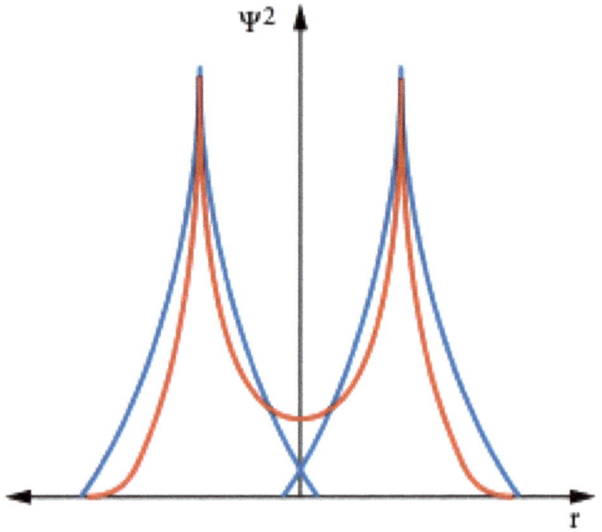

Figure 10.2. Comparison of the σ_{1s} orbital with two H_{1s} orbitals. The atoms are located at the peak of the distributions.

density between the nuclei at the expense of regions outside of that area. Note also that the σ bonding orbital decreases much faster outside of the "bond".

H_2^+ has only one electron, a rather strange bond, but Coulomb's law still applies and the potential energy is given by:

$$V = -\left(\frac{e^2}{r_A} + \frac{e^2}{r_B}\right) + \frac{e^2}{r_{AB}} \tag{10.1}$$

The first term is electron-nucleus attraction for each nucleus, while the second is nuclear-nuclear repulsion. The potential energy is lowest when the electron is near one of the nuclei or close to both of them. We conclude that bond formation permits the electron to be located in a large region of space for which the energy is low. The total energy is lowered by forming the bond and we have found the rationale for bonding! Occupation of an orbital such as that shown above in Figure 10.1 leads to a stable bond. It is called a **bonding orbital**. The ground electronic state of H_2^+ has the single electron in this orbital.

Let's go back to our thought experiment. We can create the first excited state of H_2^+ the following way. Look at the probability density plot shown in Figure 10.3.

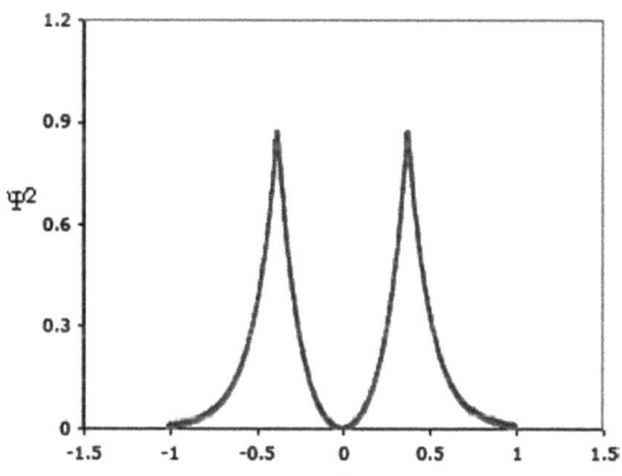

Figure 10.3. σ^*_{1s} orbital for H_2^+.

Now, the probability density between H_A and H_B is very small, almost zero. So the electron is most likely to be found somewhere other than between the nuclei. This orbital does not lead to a stable bond and we expect dissociation to $H + H^+$. This is an **antibonding orbital.** We label this antibonding orbital σ_{1s}^* and the bonding orbital σ_{1s}.

The question you should now be asking is whether or not atomic orbitals have any relationship to molecular orbitals. In fact, we can form MOs from a **L**inear **C**ombination of **A**tomic **O**rbitals, MO-LCAO. Mathematically, the MO would be expressed as:

$$\begin{aligned} \Psi_{\sigma_{1s}} &\propto \Psi_A(1s) + \Psi_B(1s) \\ \Psi_{\sigma_{1s}^*} &\propto \Psi_A(1s) - \Psi_B(1s) \end{aligned} \tag{10.2}$$

where only normalization constants are missing. Figures 10.1 to 10.3 demonstrate this process. We may also look at these linear combinations using atomic orbital cloud representations as shown in Figure 10.4.

The *Pauli Exclusion Principle* (two electrons per orbital), *Hund's Rule* (put one electron into each orbital of the same energy before pairing electrons) and the *Aufbau Principle* (fill orbitals in order of increasing energy) all apply to MOs. We can draw a MO energy level diagram to correlate the atomic and molecular energy levels for H_2^+ as shown in Figure 10.5.

We write the H_2^+ electron configuration as σ_{1s}. Note the analogy to the hydrogen atom ground state electron configuration, denoted as $1s$. Let's look at the electron configurations of the four simplest molecules. That is, look at those formed from atoms having only $1s$ electrons.

$$H_2^+ : \sigma_{1s}$$

$$He_2^+ : \sigma_{1s}^2\sigma_{1s}^*$$

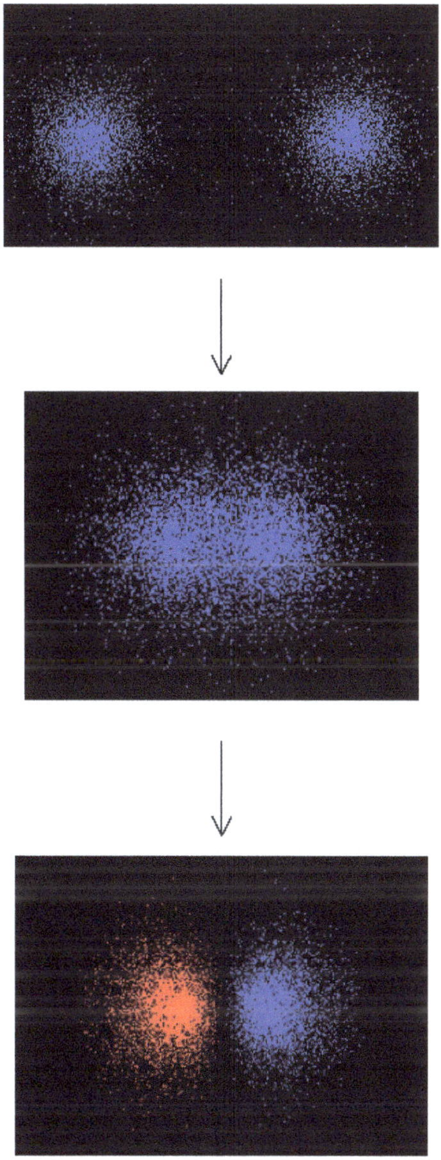

Figure 10.4. Schematic view of the formation of σ_{1s} and σ^*_{1s} orbitals from H_{1s} orbitals. Two separated $1s$ orbitals are shown in the top panel. If the two orbitals combine in phase (add), the bonding orbital in the middle panel is obtained. If they combine out of phase (subtract), the antibonding orbital shown in the bottom panel results.

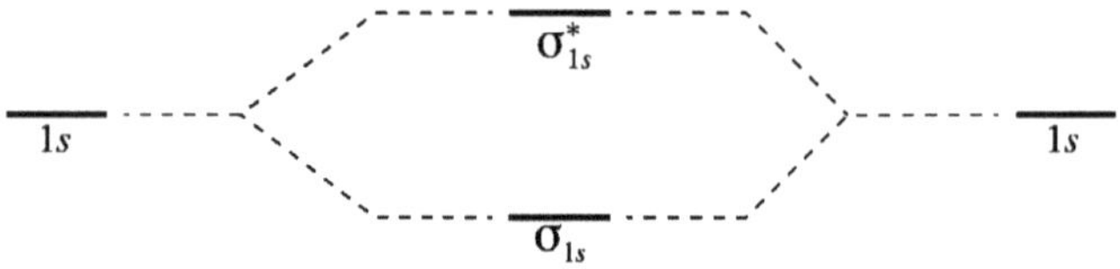

Figure 10.5. Lowest-energy MOs for molecules with four or fewer electrons.

$$H_2 : \sigma_{1s}^2$$

$$He_2 : \sigma_{1s}^2 \sigma_{1s}^{2*}$$

The MO electron configurations allow the following definition of bond order. Bond order (BO) is defined as:

$$\begin{aligned} BO = &\ {}^1\!/_2\ (\textit{number of bonding electrons}) \\ &- {}^1\!/_2\ (\textit{number of antibonding electrons}) \end{aligned}$$

If BO = 1, we have a single bond, etc. Now we can write bond orders for these molecules and correlate the bond orders with experimental bond energies (BE).

$$H_2^+ \text{ has BO} = {}^1\!/_2 \quad \text{and} \quad \text{BE} \sim 270\,\text{kJ mole}^{-1}$$

$$H_2 \text{ has BO} = 1 \quad \text{and} \quad \text{BE} \sim 450\,\text{kJ mole}^{-1}$$

For He_2^+ and He_2, the results are BO = ½, BE =∼ 310 kcal mole^{-1} and BO = 0, BE = 0, respectively.

Bond energy is the energy required to break the bond. A larger bond energy implies a more stable molecule. Therefore, we can make the following statements. H_2 is the most stable of these molecules because it has the highest bond energy/bond order. He_2 has a bond order of zero, so it will not exist. (Note that recent experimental results indicate that He_2 is very weakly bound, with a "bond energy" less than 1 kJ mole^{-1} and an effective "bond length" of 15 Å).

The next molecules to explore are Li_2, Be_2, B_2, etc. Clearly, we need additional AOs for the construction of new MOs. The

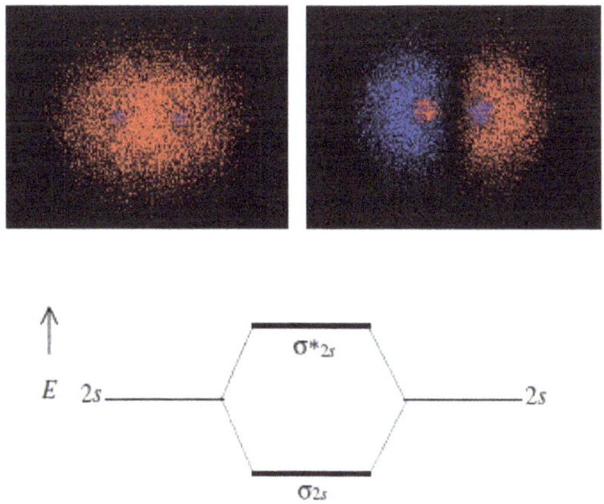

Figure 10.6. LCAO formation of σ_{2s} and σ^*_{2s} orbitals (top) and the corresponding energy level diagram (bottom).

next AO is the 2s orbital. We might guess that we can form MOs as we did with 1s orbitals and that would be correct as shown in Figure 10.6.

Since the 2s orbital is higher in energy than the 1s, we expect that the σ_{2s} and σ^*_{2s} will be higher in energy than the σ_{1s} and σ^*_{1s}. Again, our expectations are valid.

Let's keep adding AOs to our scheme. The next AO is 2p. This orbital is not spherically symmetric, so the MOs must have a different shape. We need to define an axis for the nuclei. Let's follow the convention and assume that the nuclei lie on the z axis. Begin with the p_z orbital, since in this orbital lies on the internuclear axis.

$$\Psi^*_{\sigma 2pz} \propto \Psi_A(2p_z) + \Psi_B(2p_z)$$

The combination, as shown in Figure 10.7, has a decrease in probability density between the nuclei. We do not get a bonding orbital! The sum of two p_z orbitals is $\sigma^*_{2p_z}$. Now, try subtracting two p_z orbitals.

$$\Psi_{\sigma_{2p_z}} \propto \Psi_A(2p_z) - \Psi_B(2p_z)$$

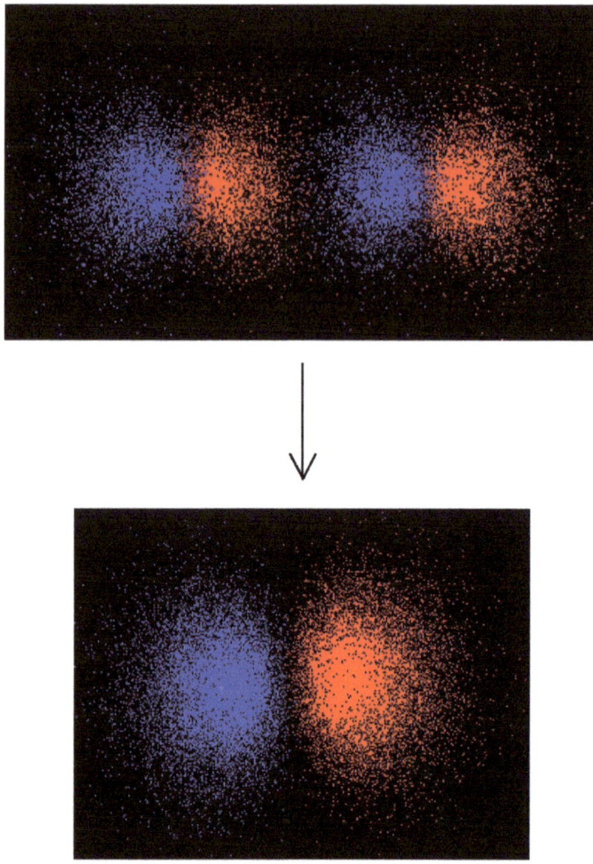

$$\Psi_{\sigma^*_{2p_z}} \propto \Psi_{2p_z}(A) + \Psi_{2p_z}(B)$$

Figure 10.7. LCAO formation of the $\sigma^*_{2p_z}$ orbital. The two p orbitals are combined in phase, but this combination corresponds to an antibonding molecular orbital.

If you are assuming that this will create a bonding orbital, Figure 10.8 indicates you are correct. This is the opposite of s-based σ orbitals!

We still must combine the remaining p orbitals into MOs. Since p orbitals have directionality, we must combine p_x with p_x and p_y with p_y. (This is because p_x has a nodal plane in the

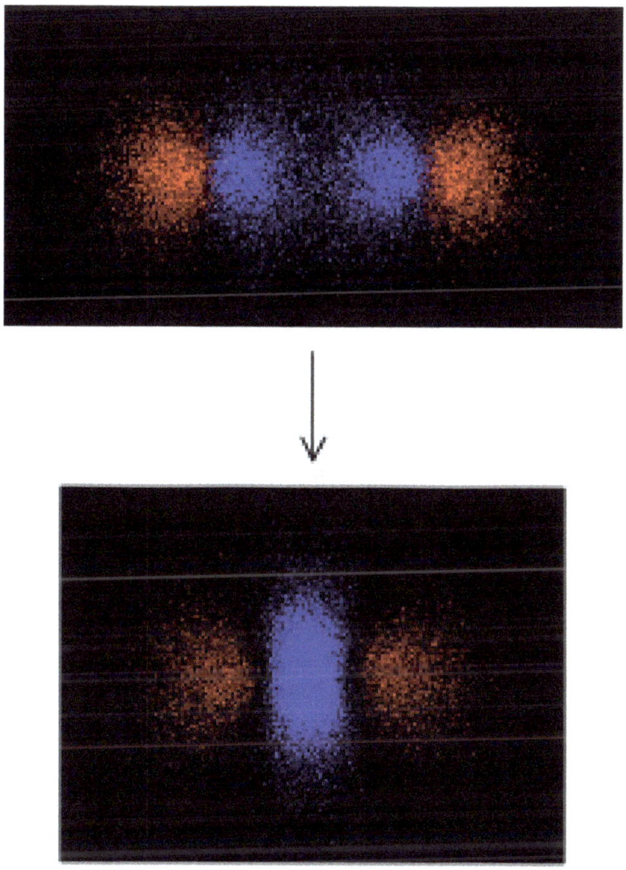

$$\Psi_{\sigma_{2p_x}} \propto \Psi_{2p_x}(A) - \Psi_{2p_x}(B)$$

Figure 10.8. LCAO formation of σ_{2p} orbital. The two p orbitals are combined out of phase to produce a bonding MO.

y plane and vice versa). Let's look at the combinations of p_x orbitals and then we can generalize.

Note that in this MO, shown in Figures 10.9 and 10.10, we have increased the probability density between the nuclei, but off the internuclear axis. This is a π-bonding orbital. We can surely suggest what will happen if we subtract two p_x orbitals.

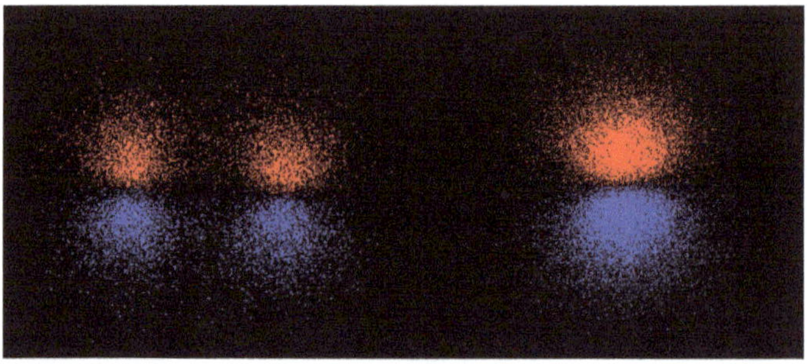

Figure 10.9. LCAO formation of π_{2p} orbital from two p_x or two p_y orbitals combined in phase.

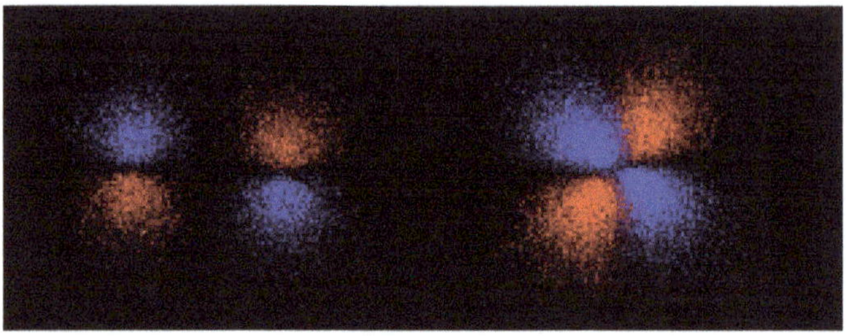

Figure 10.10. LCAO formation of the π^*_{2p} orbital from two p_x or two p_y atomic orbitals combined out of phase.

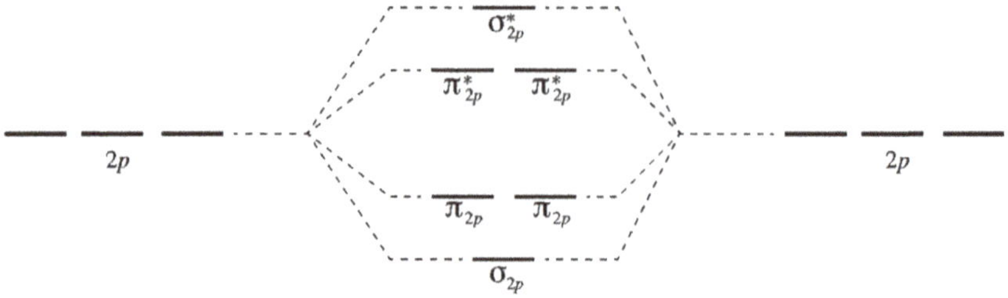

Figure 10.11. Proposed energy level diagram for MOs formed from *2p* orbitals.

We will get a decrease in probability density between the nuclei and off the axis, a π^* orbital.

Now, we can combine all of this into an energy level diagram for p orbital-based MOs as shown in Figure 10.11.

This is a good time to review what we know so far about the LCAO-MO method. The energy of a particular MO is dependent upon:

(1) the nature of the AOs used to form the MOs. Since 1*s* orbitals are of lower energy than 2*s* orbitals, MOs formed from 1*s* AOs will be lower in energy;
(2) the type of MO. Bonding orbitals are lower in energy than antibonding orbitals formed from the same AOs; and
(3) the effectiveness of the AO overlap in the MO. Generally, *s* orbitals overlap to a greater extent than *p* orbitals. However, other energetic factors may influence the relative ordering of σ and π_{2p} orbitals. We explore that in more detail below.

For heavy atoms O and F, the 2*s* and 2*p* orbitals are very different in energy and the σ_{2p} orbital is filled before the π_{2p} orbital. The opposite is true for the atoms Li through N. The energy levels are shown in Figure 10.12.

The change in ordering for the light atoms is due to a repulsive interaction between σ_{2s} and σ_{2p}, which occurs because the atomic *s* and *p* orbitals are close in energy.

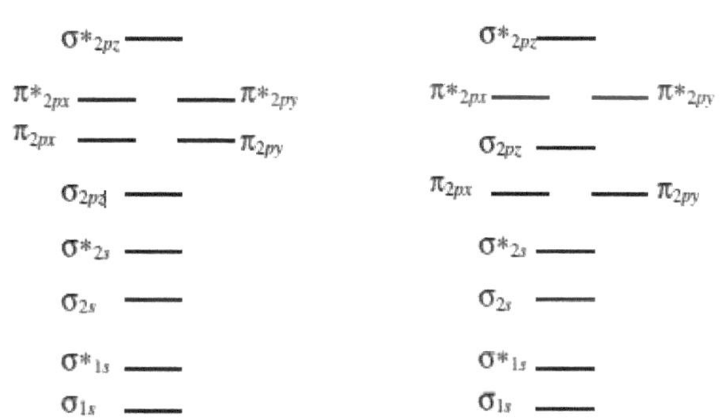

Figure 10.12. Energy level diagrams for MOs from atomic orbitals through 2*p*. The ordering for both heavy (left) and light atoms (right) is shown.

Examples

Let's look at the success of MO theory by examining the electronic structure of two diatomic molecules, O_2 and N_2. We note to begin that O_2 is paramagnetic (that is, it exhibits magnetism due to unpaired electrons) and N_2 is diamagnetic (it does not exhibit magnetism, indicating that all of its electrons must be paired.)

Oxygen is a heavy atom, so that the electron configuration of O_2 is $\sigma_{1s}^2\sigma_{1s}^{*2}\sigma_{2s}^2\sigma_{2s}^{*2}\sigma_{2p}^2\pi_{p_x}^2\pi_{p_y}^2\pi_{p_x}^*\pi_{p_y}^*$ and the bond order is 2. We observed that there are two half-filled orbitals and the molecule will be paramagnetic.

Nitrogen is a light atom, so that the electron configuration of N_2 is $\sigma_{1s}^2\sigma_{1s}^{*2}\sigma_{2s}^2\sigma_{2s}^{*2}\pi_{p_x}^2\pi_{p_y}^2\sigma_{2p}^2$ and the bond order is 3. All of the orbitals have paired electrons and the molecule is expected to be diamagnetic.

So, we conclude from this simple analysis that we are in a good place with MO theory. An interesting classroom demonstration that MO theory is correct is to pour liquid nitrogen and liquid oxygen through a magnet. Liquid N_2 flows directly through the magnet, but liquid O_2 is captured between the two poles of the magnet.

We have restricted all of this to homonuclear diatomic molecules. These are obviously a very small subset of the possible diatomic molecules. It is time to move on to heteronuclear molecules. We already know what needs to be considered. Let's write some configurations first, then look at the MOs in detail. Note that the presence of even one "heavy" atom requires the use of the heavy atom-derived MO energy level diagram.

Molecule	Configuration	Bond Order
CN (13 electrons)	$\sigma_{1s}^2\sigma_{1s}^{*2}\sigma_{2x}^2\sigma_{2x}^{*2}\pi_{2px}^4\sigma_{2p}$	2.5
CO (14 electrons)	$\sigma_{1s}^2\sigma_{1s}^{*2}\sigma_{2x}^2\sigma_{2x}^{*2}\sigma_{2p}^2\pi_{2p}^4$	3
NO (15 electrons)	$\sigma_{1s}^2\sigma_{1s}^{*2}\sigma_{2x}^2\sigma_{2x}^{*2}\sigma_{2p}^2\pi_{2p}^4\pi_{2p}^*$	2.5

We need to compare heteronuclear and homonuclear results. Are N_2 and CO really the same, since they are isoelectronic? We can make this comparison using probability densities as in Figure 10.13.

$$\Psi_{\sigma_{1s}} = c_A \Psi_{1s}(N) + c_B \Psi_{1s}(N) \tag{10.5}$$

$$\Psi_{\sigma_{1s}} = c_A \Psi_{1s}(C) + c_B \Psi_{1s}(O) \tag{10.6}$$

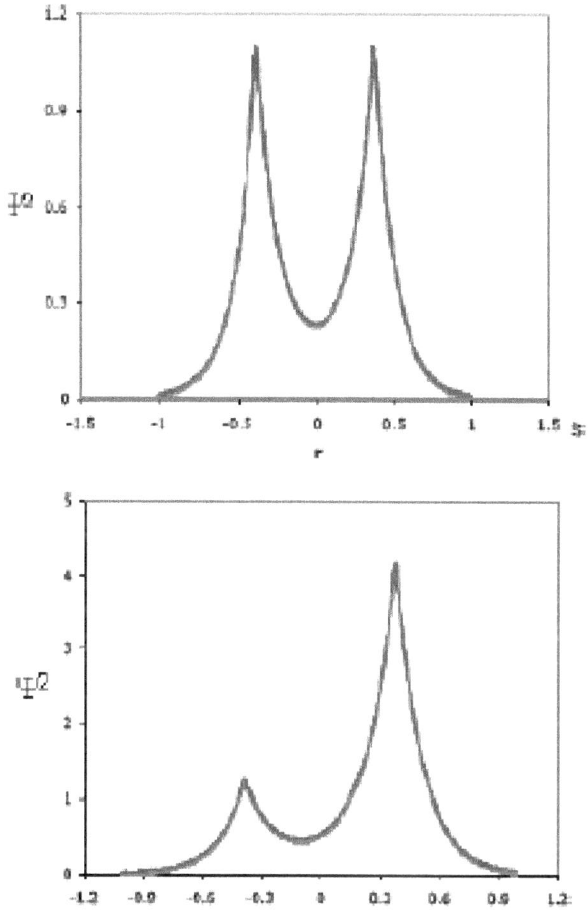

Figure 10.13. LCAO formation of σ_{1s} orbitals for N_2 (top) and CO (bottom).

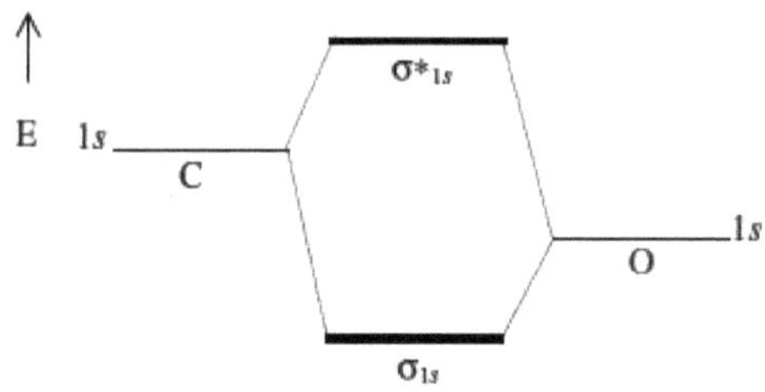

Figure 10.14. CO molecular orbital energies for σ_{1s} and σ^*_{1s} orbitals.

The asymmetry in the CO probability density is due to the difference in AO energies and correlates with electronegativity; probability density is greater near the more electronegative atom. The MO energy diagram, shown in Figure 10.14, reflects this difference. The bonding MO energy is closer to the oxygen AO energy, while antibonding MO energy is closer to the carbon AO energy.

Let's look at an extreme example of this asymmetry, the nearly ionic molecule HF. The first step is to determine the orbital energies so that we can decide which AOs will combine into MOs.

$$\mathrm{H}_{1s}:\ -1310\,\mathrm{kJ\,mole}^{-1} \qquad \mathrm{F}_{1s}:\ -42000\,\mathrm{kJ\,mole}^{-1}$$
$$\mathrm{F}_{2s}:\ -3400\,\mathrm{kJ\,mole}^{-1}$$
$$\mathrm{F}_{2p}:\ -1400\,\mathrm{kJ\,mole}^{-1}$$

F_{1s} will not combine with H_{1s} because energy of the fluorine orbital is so low. An energy mismatch also occurs between the F_{2s} and H_{1s} orbitals. The bond must form by combining the H_{1s} and F_{2p} orbitals.

To produce a σ orbital, we can only combine F_{2p} with H_{1s}, with resultant energy level diagram shown in Figure 10.15.

HF has a total of 10 electrons. Using the Pauli Exclusion Principle and the Aufbau Principle, fill the orbitals with electrons starting from the lowest orbital. The resulting electron

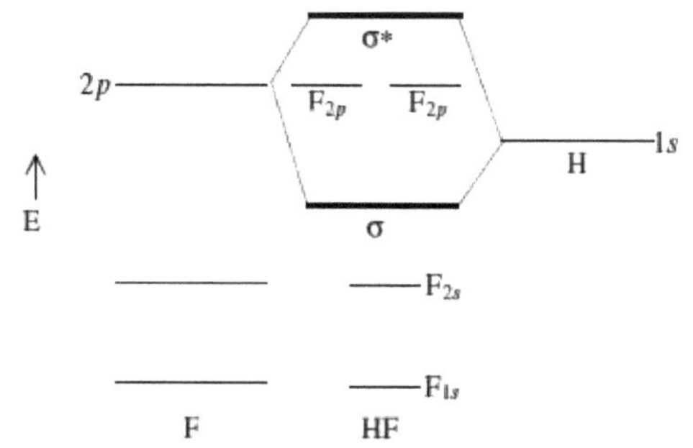

Figure 10.15. HF molecular orbital energies.

configuration is

$$F_{1s}^2 F_{2s}^2 \sigma^2 F_{2p_x}^2 F_{2p_x}^2$$

The bond order is one and the F_{1s}, F_{2s}, F_{2p} orbitals are all nonbonding orbitals (that is, neither bonding nor anti-bonding).

To review, we have found our successful bonding model. MO theory provides us with the important rationale for bonding. It directly provides a numerical result for the lowering of the energy that occurs when bonds form from atoms. It also provides structural details for the molecule. MO theory has proven quite successful and is the most widely used model for the formation of molecules.

Exercises

1. What are the electron configurations and bond orders for Be_2, Be_2^+ and Be_2^-?
2. What are the electron configurations and bond orders for the two series of ions CO^+, CO^-, CN^+ and CN^-? Compare these ionic configurations to the neutral CO and CN molecules, respectively. Are any of these six ions/molecules isoelectronic?

3. Compare O_2 and its ions, O_2^+ and O_2^- to the electron configurations and bond orders for N_2 and its ions.
4. Construct an MO energy level diagram for the LiH molecule. The relevant AO energy levels are: Li(1s) $-7300\,\text{kJ mole}^{-1}$; Li(2$s$) $-520\,\text{kJ mole}^{-1}$; H(1$s$) $-1312\,\text{kJ mole}^{-1}$.

CHAPTER 11

Valence Bond Theory

We have listed this as a separate bonding model, but in reality this is not a different theory. We can make valence bond theory take the same mathematical form as MO theory, if we wish. However, this defeats its major benefit: a physical picture of bonding in large (especially organic) molecules. As always, we begin with something simple. In this case, we start with BeH_2. We already know that the H–Be–H bond angle is 180°. Be has 2 valence electrons, a $1s^22s^2$ configuration. Since all of the electrons are paired, we can only imagine forming two bonds (actually, we can't imagine forming any bonds without changing the configuration) by unpairing the $2s$ electrons, so that the electron configuration becomes $1s^22s2p$. But how can this help? There are two problems:

(i) $2p$ electrons are higher in energy than $2s$, so how can we justify unpairing the electrons?
(ii) How can we get two equivalent bonds from a $1s^22s2p$ electron configuration?

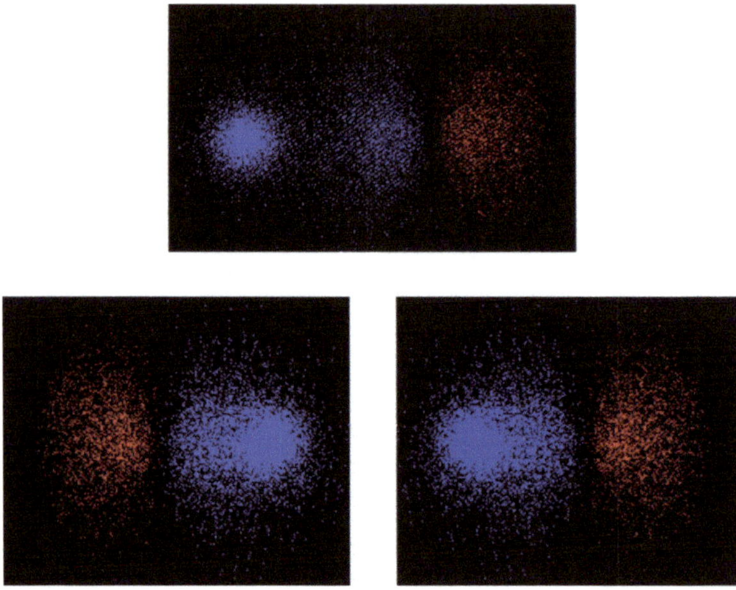

Figure 11.1. Schematic for formation of *sp* hybrid orbitals in Be. The top panel shows the *s* and *p* orbitals at a large distance apart and the bottom panel shows the two hybrid orbitals formed by the sum and difference of the two atomic orbitals. They point 180° apart.

The solution to both of these difficulties is to use *hybrid atomic orbitals*. In this case, *sp* hybrid orbitals. It is best to view these pictorially (Figure 11.1).

The two *sp* orbitals then combine with two H(1s) orbitals to form two σ bonds, 180° apart. The two σ bonds will be equivalent since each consists of an overlap between a Be(sp) orbital and a H(1s) orbital.

Hybrid atomic orbitals are mathematical combinations of the "normal" atomic orbitals. They are as "real" as hydrogen-like atomic orbitals, i.e., they are mathematical constructs. Hybrid orbitals are an effective way to describe multi-electron/multi-atom systems, i.e., molecules. They are extremely useful in organic chemistry since the molecules are too large to easily and transparently treat with MO theory and easily provide a physical picture.

Let's look at a more complex molecule and explore the energetics of the bonding process. We'll do a thought experiment and construct a mechanism for the formation of CH_4. Remember, it does not actually form this way. This is only an attempt to rationalize observations.

Our mechanism:

(i) start with ground state carbon ($1s^2 2s^2 2p^2$) and four ground state hydrogens (1s);
(ii) promote a carbon $2s$ electron to the $2p$ orbital so that the configuration becomes $1s^2 2s 2p^3$ and we have four half-filled orbitals;
(iii) use the $2s2p^3$ electrons to form 4 sp^3 hybrid orbitals;
(iv) form CH_4 from this configuration.

The process is shown on the energy level diagram in Figure 11.2.

In our "mechanism", we must have an energy input of 406 kJ mole^{-1}, but the hybridization process is cost-free in terms of energy. The endoergic process of forming sp^3 orbitals leads to formation of four C–H bonds and an energy lowering of 2,068 kJ mole^{-1}. The net energy change for the overall "mechanism" is more exoergic than it would be if we simply formed CH_2 from the ground state carbon atom. In the carbon molecule, the

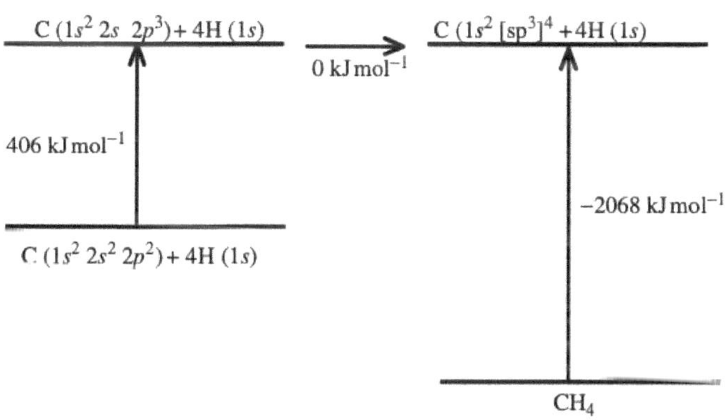

Figure 11.2. Schematic mechanism for formation of CH_4.

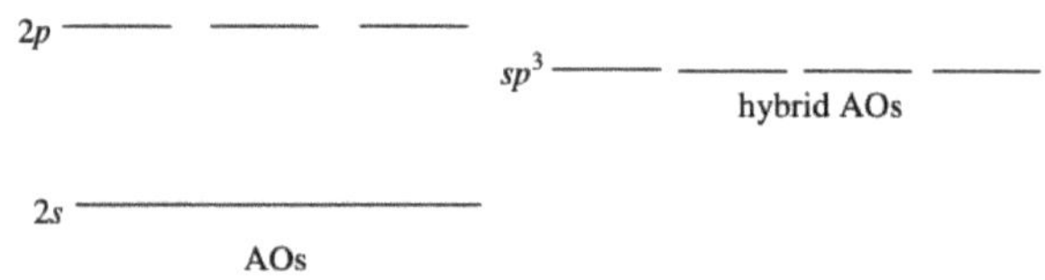

Figure 11.3. Relative ordering of $2s/2p$ atomic and sp^3 hybrid orbitals.

hybrid orbital is $^1/_4$ s and $^3/_4$ p. The energy of the sp^3 orbital is $^3/_4$ of the energy of the $2p$. You can see this energy scaling in Figure 11.3.

In the end, for simple molecules, much like VSEPR, we have an orbital hybridization scheme that is based solely on the number of bonds. (We treat non-bonding pairs of electrons as "bonding pairs" and multiple bonds as one bonding pair, just as in VSEPR). The fundamental difference in comparison to VSEPR theory is that we now have a theoretical basis (energy lowering) for the assignment of geometry.

σ Bonding Electron Pairs	**Hybrid Orbitals**	**Ideal Geometry**
2	sp	linear
3	sp^2	trigonal
4	sp^3	tetrahedral
5	dsp^3	trigonal bipyramidal
6	d^2sp^3	octahedral

Examples of Valence Bond Model Application

(1) Nitrogen, N_2

- Begin with Lewis dot structure :N≡N:.
- Each atom has one σ bond plus one non-bonding pair of electrons, so the hybrid orbitals will be sp.
- The original electron configuration is $1s^22s^22p^3$.

- The hybrid orbital configuration is $1s^2(sp)^3 2p_x 2p_y$ for each N atom.
- One of the *sp* hybrid orbitals is filled with a non-bonding pair of electrons and the second overlaps the *sp* orbital from the other N to form a σ bond.
- The two π bonds are formed by the overlap of p_x orbitals from each of the two N atoms and the overlap of p_y orbitals from each atom.

(2) Acetylene, C_2H_2 (H–C≡C–H)

- Each C has two "bonding pairs" by our definition. Therefore, we expect *sp* hybrid orbitals in the molecule for each C.
- The orbital configuration becomes $C(1s^2(sp)^2 p_x p_y)$.
- σ bonds form by overlap of a H(1*s*) with an *sp* orbital on each C and between the two C atoms using overlap of an *sp* orbital from each C.

$$\left.\begin{array}{r} C_{sp}^{(A)} - C_{sp}^{(B)} \\ C_{sp}^{(A)} - H \\ C_{sp}^{(B)} - H \end{array}\right\} \sigma \text{ bonds}$$

$$\left.\begin{array}{r} p_{2p_x}^{(A)} - p_{2p_x}^{(B)} \\ p_{2p_y}^{(A)} - p_{2p_y}^{(B)} \end{array}\right\} \pi \text{ bonds}$$

- The π orbitals lock in the geometry. It is not possible to rotate the carbon atoms and maintain the orientation of the *p* orbitals needed for the two π bonds. The π bonds are oriented 90° apart

(3) Ethylene ($H_2C{=}CH_2$)

- There are three pairs of electrons around each C atom. These are used in the σ bonding framework.

- The carbon hybrid electron configuration is $1s^2(sp^2)^3p_x$.

$$\left.\begin{array}{cc} C^{(A)}_{sp^2} - H(1s) & C^{(B)}_{sp^2} - H(1s) \\ C^{(A)}_{sp^2} - H(1s) & C^{(B)}_{sp^2} - H(1s) \\ C^{(A)}_{sp^2} - C^{(B)}_{sp^2} & \end{array}\right\} \sigma \text{ bonds}$$

$$\left.C^{(A)}_{P_x} - C^{(B)}_{P_x}\right\} \pi \text{ bond}$$

Again, the π orbital locks in the geometry. Since p_x must overlap with p_x, we get a planar geometry. Ethane (C_2H_6) has no π bonds and may freely rotate. Ethane involves C with sp^3 hybrid orbitals.

Exercises

1. Describe the bonding in BeH_3 using valence bond theory.
2. The organic molecule 1,3-butadiene ($H_2C{=}CH{-}CH{=}CH_2$) is structurally related (roughly) to ethylene. Describe the bonding around each carbon atom.
3. Now, consider the cyclic form of the molecule in Exercise 3: cyclobutadiene shown below with 90° C–C–C angles. Can you make any comment on the stability of this molecule by considering valence bond theory?

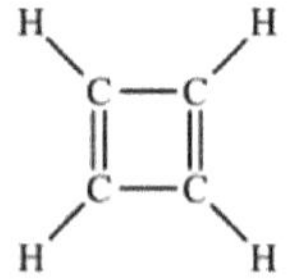

4. We want to compare valence bond theory and MO theory descriptions of bonding. MO theory was used to describe O_2 bonding in Chapter 10. Compare that result to the result from valence bond theory.

CHAPTER 12

Other Types of Bonding

Hydrogen bonds occur only in the most polar molecules. That is, those containing the elements with the largest electronegativity: N, O and F. These are not bonds in the sense that we have been using that term.

Hydrogen has only a single electron involved in bonding. It can closely approach an electronegative atom without any electron-electron repulsion. This is the interaction we label as a hydrogen bond. The bonds are a result of the skewed electron distribution in these molecules. One end is essentially positive and the other is essentially negative. For example, look at HF, where dimers are dominant.

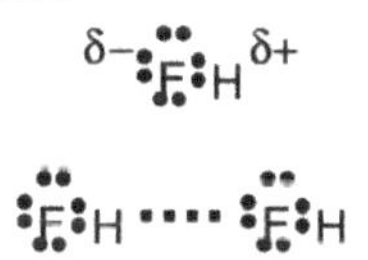

Hydrogen bonds occur in all phases: solid, liquids and gases. Typical hydrogen bond energies are less than 60 kJ $mole^{-1}$

and bond lengths are approximately 2Å. Compare those values with a typical covalent bond where the bond energy is 400–500 kJ mole^{-1} and bond lengths are 1.0–1.3 Å. Hydrogen bonding is most evident in the determination of boiling points of hydrogen-containing molecules. Specifically, compare the hydrides of Groups IV, VI and VII.

CH_4	−170°C	H_2O	100°C	HF	20°C
SiH_4	−110°C	H_2S	−60°C	HCl	−90°C
GeH_4	−95°C	H_2Se	−40°C	HBr	−70°C
SnH_4	−50°C	H_2Te	0°C	HI	−40°C

Group IV is our control. As one goes down the periodic table, boiling point increases. Groups VI and VII exhibit the same trend, except for H_2O and HF. These are hydrogen-bonded molecules. Without hydrogen bonding, we predict H_2O would boil at approximately −80°C! The need to "break" hydrogen bonds contributes to the enthalpy of vaporization and causes an increase in boiling point. Hydrogen bonding leads to order in liquids and solids. For example, we have the "Mickey Mouse" model of hydrogen bonding in H_2O as shown in Figure 12.1.

Similar, but weaker, interactions occur between other atoms and molecules. What are the forces that cause molecules to be attracted to each other? How are liquids "held" together? The answer is that these processes occur via intermolecular forces — they are much weaker than covalent, ionic bonds or even hydrogen bonds. We'll look at a series of interactions of varying strength and apply the results to experimental observations.

The first type is a general class called *van der Waals forces.* There are actually several specific types of such interactions. They are responsible for deviations from ideal gas behavior at high temperatures and low pressures and also for condensation of gases at sufficiently low temperatures.

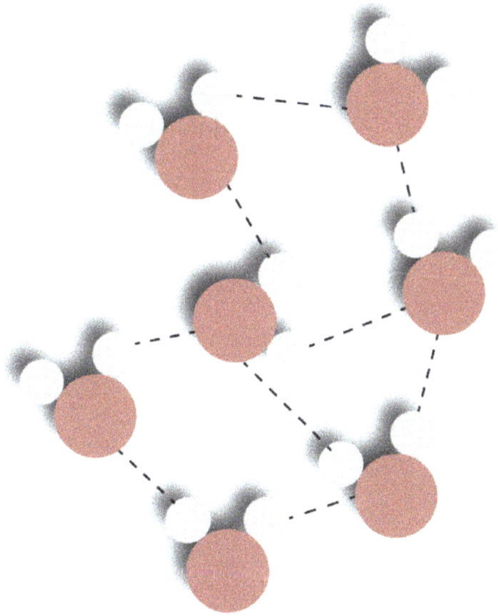

Figure 12.1. Schematic representation of the hydrogen-bonded structure of liquid water.

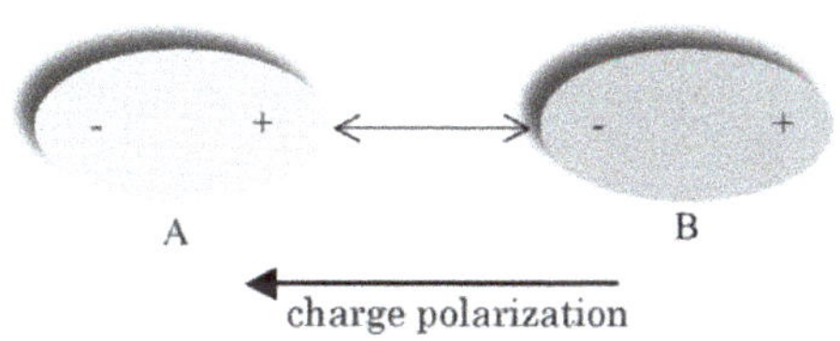

Figure 12.2. Schematic representation of induced polarization.

(1) The *London dispersion force* is depicted in Figure 12.2. A momentary fluctuation in the charge distribution of A induces charge polarization on B and then there is an electrostatic attraction. The strength of the interaction is given by:

$$V_L = -\frac{A}{r^6} \tag{12.1}$$

Compare this result with Coulombic forces in ionic bonds, which scale as $1/r$. London forces are operative over a very small range of r. The constant A includes the polarizability of the molecule, which is a measure of the extent of distortion of the electron cloud of an atom by the electric field of nearby atoms. Larger atoms generally have the highest polarizability because electrons far from the nucleus are more loosely held. Shapes of molecules also affect polarizability. Spherical molecules are less polarizable than elongated molecules.

(2) *Dipole-dipole interactions* are stronger than dispersion forces. The drawing in Figure 12.3 provides a visual clue to the nature of these interactions. If two molecules are polar and come together, there is an attractive interaction (provided that the alignment is correct). The potential energy has the same form as in dispersion forces, but the constant A now depends on the dipole moment rather than the polarizability.

(3) *Repulsive interactions* are operative only at very small intermolecular separations. When molecules approach too closely, electrons in filled atomic orbitals begin to overlap and repel one another. The potential energy is given by:

$$V_{rep} = \frac{B}{r^{12}} \tag{12.2}$$

We can combine the London attractive term with this repulsive term to construct a realistic potential energy function known as

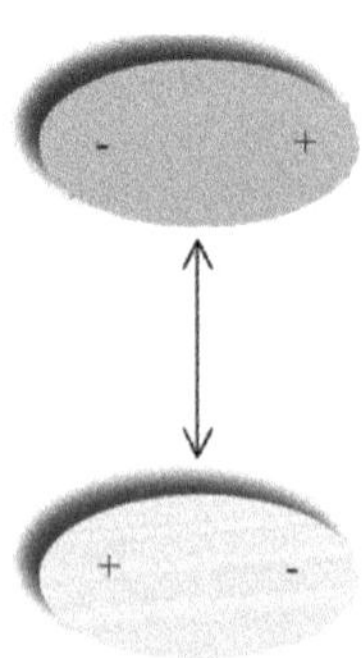

Figure 12.3. Schematic representation of dipole-dipole interactions.

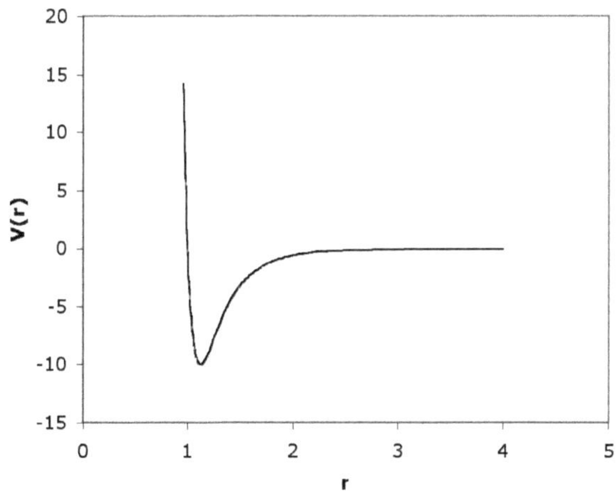

Figure 12.4. The generic Lennard-Jones potential function.

the *Lennard-Jones potential*, shown in Figure 12.4.

$$V_{LJ} = -\frac{A}{r^6} + \frac{B}{r^{12}} = 4\epsilon\left[\left(\frac{\sigma}{r}\right)^{12} - \left(\frac{\sigma}{r}\right)^{6}\right]$$

The constant ε is the minimum potential energy and σ is the separation at which the potential energy becomes positive. The values of ε are the formal equivalents of bond energies. However, a bond energy is typically 430 kJ mole^{-1} (say for H_2) while ε is 0.3 kJ mole^{-1} (for H_2-H_2). The values of σ correspond to the sum of the van der Waals radii of the atoms. We "measure" these effects by determining the melting and boiling points.

Intermolecular forces also have macroscopic manifestations. Now, we can take a more macroscopic look at these forces. What do we see when we look at liquids or solids rather than individual pairs of atoms or molecules as in the gas phase? In liquids, we find macroscopic properties such as low compressibility, non-rigidity and high density.

Why do liquids appear as droplets? Let's look at Figure 12.5. Focus on the dark molecule at the center of this droplet. A molecule in the interior experiences a number of attractive interactions from all sides, while the same molecule at the surface has

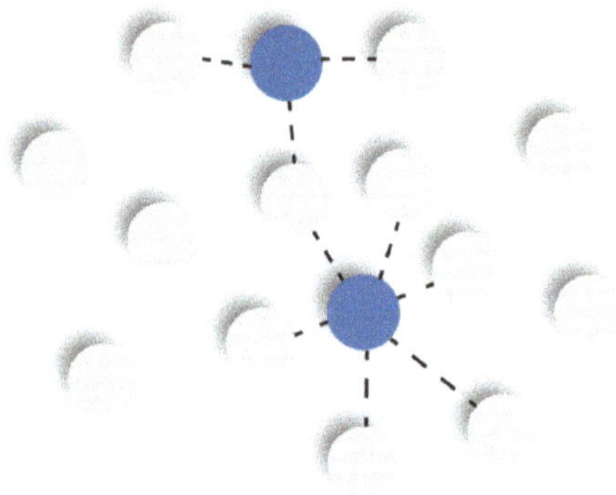

Figure 12.5. Intermolecular forces and the macroscopic effect of surface tension.

fewer possible interactions. The net effect is to pull molecules toward the center, leaving the droplet with a minimum of surface area, that is, a spherical shape. To increase surface area, molecules must move from the interior to the surface. This requires energy since an intermolecular force must be overcome. The resistance to the increase in surface area is *surface tension.* The larger the intermolecular forces, the higher the surface tension. These are *cohesive* forces.

Liquids exhibit capillary action. This is the spontaneous rising of a liquid in a small diameter tube. If the molecules of liquid are polar and the tube has a surface with polar bonds (like glass), we observe *adhesive forces* (glass has dangling oxygen atoms with partial negative charges, which attract the positive end of the H_2O dipole). Water "climbs up" the tube at the walls, but cohesive forces try to pull H_2O away from the walls. There is a balancing effect and we see a *meniscus.* Note that there are two possible shapes as shown in Figure 12.6. A *concave meniscus* occurs in a liquid in which the adhesive forces are greater than the cohesive forces. The liquid climbs to a height where the weight of liquid column balances the attraction to the walls. A *convex meniscus* occurs in a liquid in which the cohesive forces

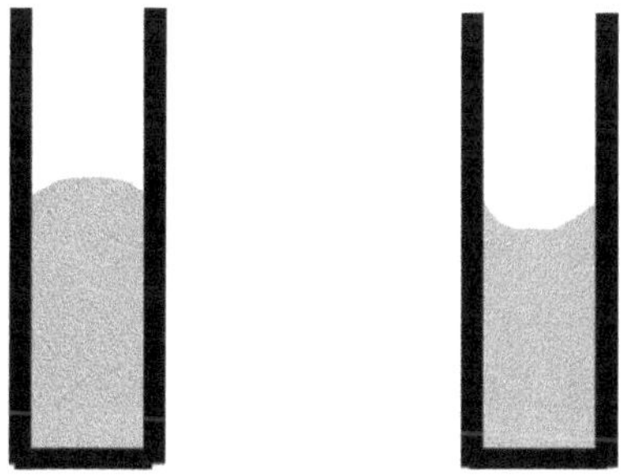

Figure 12.6. A convex (left) meniscus and a concave (right) meniscus.

are greater than the adhesive forces (an example is mercury in a glass tube).

Viscosity is another macroscopic manifestation of intermolecular forces. This is the fluid resistance to flow. If the forces are large, then the viscosity is high. If a molecule forms hydrogen bonds, then it will be viscous.

Exercises

1. Which molecule will have the highest boiling point: BeH_2, BH_3, C_2H_6 ($H_3C–CH_3$), NH_3?
2. Which molecule will exhibit the greatest polarizability and hence the highest boiling point: CH_4, C_2H_6, C_3H_8 ($H_3C–CH_2–CH_3$), C_4H_{10} ($H_3C–CH_2–CH_2–CH_3$)?
3. Rank the following in order of increasing strength of the intermolecular forces: CH_4, HF, NO, C_2H_6.
4. What is the equilibrium position of the H_2 dimer (H_2-H_2)? Use Figure 12.4.
5. What is the expected meniscus shape in a glass tube for HCl? For molten gallium?

CHAPTER 13

Case Studies: Dye Molecules and Organic Molecular Stabilities

So, we have completed our journey through basic quantum theory and the application to chemical structure. Is there some way that all of this material can be tied together in an application? Of course, and we'll include extremes: a simple quantum model (the one-dimensional particle in a box) and a very large, complex dye molecule. I'm sure that the particle in a box, especially in one dimension, seemed useless at the time we developed it. It turns out that it is an excellent model for conjugated (we'll see what that means in a moment) molecules. In short, butadiene and the dye molecules constitute *case studies* in quantum mechanics/chemistry and case studies are an excellent way to explore what we have learned in science or any other area of knowledge.

The color of most dyes is due to the interaction between visible light and the atoms (or more correctly, the electrons in the atoms) that make up the molecules. It is a dynamic interaction. Light is continuously absorbed, promoting electrons to higher stationary states. The electrons subsequently drop back into the ground state configuration and repeat the cycle. The color we see is that of the incident white light *minus* the absorbed wavelengths. Recall that light will be absorbed only when its energy corresponds to the energy difference between the ground state and some excited state of the molecule:

$$E_{\text{absorbed}} = E_{\text{excited state}} - E_{\text{ground state}} \tag{13.1}$$

If we wish to predict the absorption spectrum of a molecule, we must know the energy levels of the molecule. Sadly, the hydrogen atom is the only 'real' atomic/molecular system for which an analytic solution is known. Luckily for us, for the proper choice of molecule, some of the simpler quantum mechanical models are valid. That means we must select the molecule to fit the theory! But our purpose here is to develop a case study, so we'll accept that and apply the one-dimensional particle in a box model to a specific type of dye known as a polyene. The particle in a box model was discussed in detail in Chapter 5 along with the analogous vibrating string model. There is no need to repeat those details here. But what is a polyene and how (and why) does it fit the ideal of the square well?

A stylized polyene is shown in the drawing below. In this picture, the molecule is terminated by substituted six-membered rings. The long chain of double bonds represents a *conjugated system* and we'll see that this is our one-dimensional 'box'. To understand conjugation and the meaning of the chemist's shorthand drawing shown above, we'll look at a smaller molecule that is a fragment of the polyene shown above. This molecule is known as 1,3-butadiene.

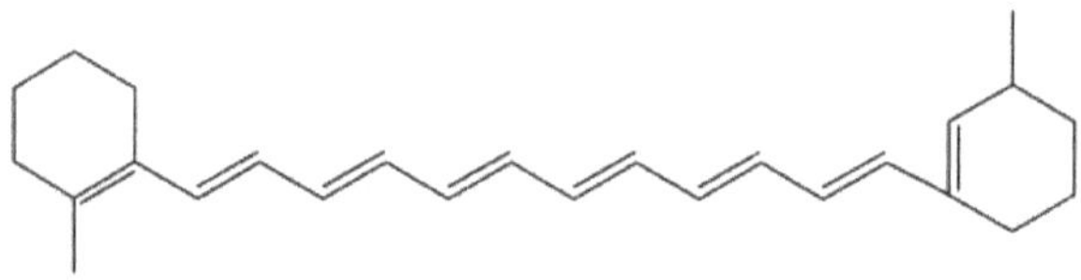

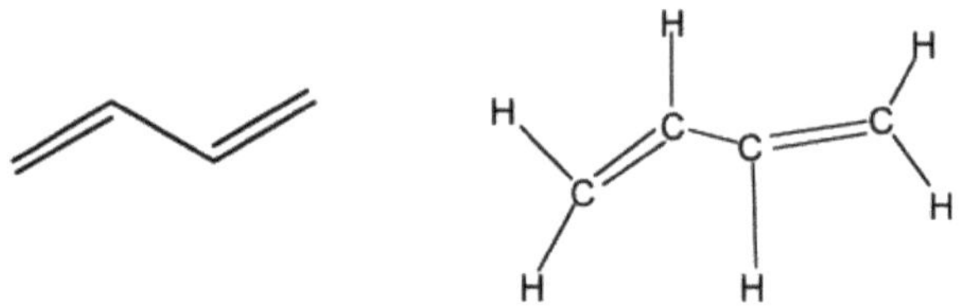

On the left is a shorthand representation as we saw for the polyene. On the right is the equivalent drawing in all of its glory. Using this conversion, you may see what symbols are assumed in the polyene drawing. The important feature in the butadiene molecule is the alternating single and double bonds. *This is conjugation.* For a better understanding of what this means, we can refer to the atomic p orbitals on the carbon atoms. Linear combinations of these orbitals make up the molecular π orbitals (Chapter 10), as shown.

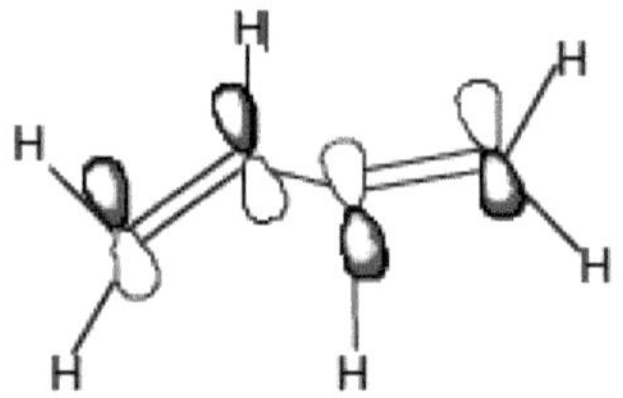

Each carbon atom contributes one electron to the π orbitals; four carbon atoms mean four π electrons. Since electrons are indistinguishable, we may assume that these four electrons are distributed over the entire four-carbon chain. Or, if we are attempting to apply a simple model, these four electrons are in a one-dimensional box with a length equal to the length of the carbon chain.

So, a polyene is simply a long chain of alternating single and double bonds, a conjugated system. In our earlier example, there are 16 π electrons, but the box length is rather undefined. Why? Resonance structures exist and the π system is not really a linear chain.

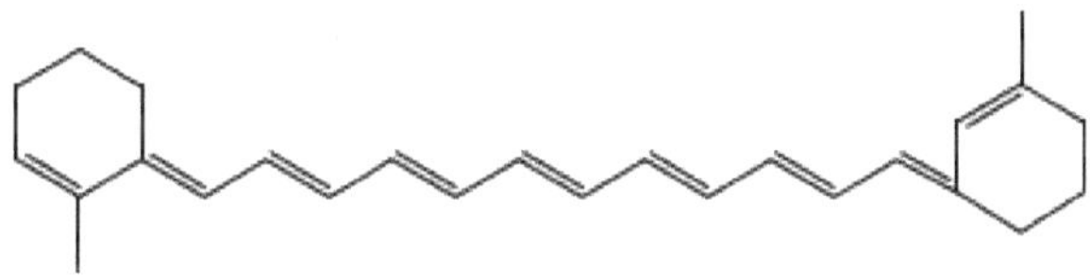

Let's begin the case study with something "real". This is an easy experiment to try. We'll use data that can be taken in the lab. The molecules are cyanine dyes. We'll use three different dye molecules with the structures imaginatively labeled A, B and C in Figure 13.1.

The only difference among these three dye molecules is the length of the carbon chain connecting the two rings. The π electron system, the box, extends from nitrogen to nitrogen across the molecules. The dyes have six, eight and ten π electrons, respectively, in the box. How big are the boxes? Time for simplifying assumptions: each bond in the chain has an average length of 0.140 nm and we will extend the box one bond beyond each nitrogen atom as did Hans Kuhn (*Journal of Chemical Physics* **17**, 1198 (1949)) who first did this type of calculation. So the boxes are 0.84 nm, 1.12 nm and 1.40 nm, respectively.

Recall the expression for the energy of an electron in a one-dimensional box and the resulting energy levels (Chapter 5):

$$E = \frac{n^2\hbar^2}{8mL^2} \quad n = 1,\, 2,\, 3,\, \ldots \tag{13.2}$$

There are two electrons per energy level, so levels up to $n = 3, 4$ and 5 are filled for dyes A, B and C, respectively. If the lowest energy transitions are considered, $n \rightarrow n + 1$, the energy (and wavelength) absorbed will be given by Equation (13.3):

$$\Delta E = E_{n+1} - E_n = \frac{(2n+1)\hbar^2}{8mL^2} \tag{13.3}$$

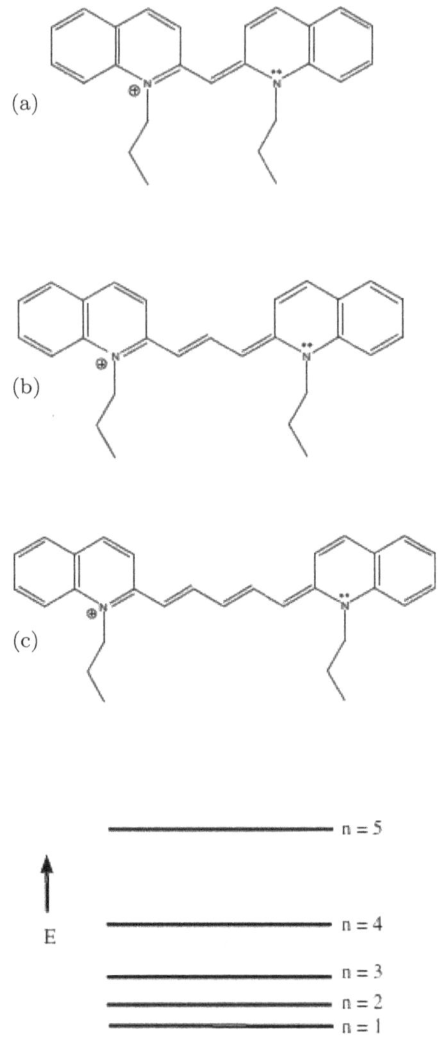

Figure 13.1. (Top) The cyanine dye molecules used in our model calculations. (Bottom) Schematic energy levels for the one-dimensional particle in a box.

Let's do the calculation of the absorbed energy for the lowest transition for each dye. We predict absorption maxima of 333 nm, 461 nm and 589 nm for dyes A, B and C, respectively. How did we do? Well, these are easy measurements to make in the lab. In fact, a number of general chemistry programs use this as a lab exercise. We have the data with which to compare. The observed maxima (each absorption is really a broad 'band' extending over more than 30 nm for reasons beyond the scope of this treatment, but we'll use the wavelength of maximum absorption) are 525 nm, 610 nm and 720 nm. Our calculated values are at lower wavelengths (higher energies) than the experimental ones, indicating that the one-dimensional boxes are effectively longer than we estimated. This agreement looks worse than it really is. Let's use the experimental results to calculate the corresponding box length. The results are 1.05 nm, 1.29 nm and 1.55 nm, respectively; the discrepancy is less startling using this metric. As we thought, the 'box' is longer than we assumed; approximately 1.5 average bond lengths longer. Why?

Faculty are always ready with explanations for the failure of simple theories, but in this case there are legitimate reasons for the discrepancies and a look at those reasons is very instructive. The model assumes that the walls of the box are perpendicular; they rise with an infinite slope. That may approach reality for a charged particle trapped in an electric field, but not for something as complex as a dye molecule. The walls are undoubtedly curved. Curved walls will increase the length of the box. In fact, Kuhn added a term to his box length expression to account for this behavior and it resulted in very good agreement with the experimental data. We could do the same, and effectively scale the results. However, it is not our intention to exactly match the experimental result. We are exploring a case study and we can see that the *relative* dye absorption wavelengths are correct. A smaller error, one that we can usually ignore, is electron-electron repulsion. We have at least six electrons in this box. Way back, we started with Coulomb's law; like charges repel. There will be a small correction due to this effect. Finally, there are 'localization' effects. The electrons 'remember' the carbon atom from which they were donated. However, this is an even smaller correction that is readily ignored.

Let's apply one of our other bonding theories to a very qualitative problem: the stability of alternative molecular structures. We saw the chain form of the 1,3-butadiene molecule earlier in the chapter. There is an alternative cyclic structure shown below. The question is a simple one: is one configuration preferred or is the "real" structure some weighted mix of these two forms?

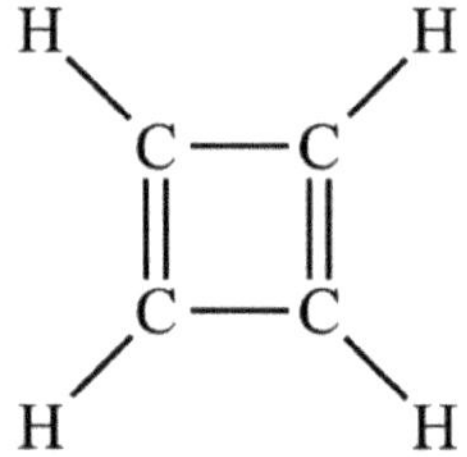

As noted, this is a qualitative problem. Each carbon atom in the chain structure is assigned sp^2 hybrid orbitals, implying a preferred 120° bond angle with the next carbon and the two hydrogen atoms. Formally, the carbon atoms in the cyclic structure are also assigned sp^2 hybrid orbitals, since the designation depends on the number of bonds. However, the bond angles in the ring are not 120°. In fact, as shown they are 90°, which would result in severe ring strain. This cyclic molecule is not expected to be stable and would be expected to ring open immediately, if you could produce it. Butadiene is expected to be exclusively a chain molecule under normal conditions.

We have applied a simple model, one that we can easily solve, to a complicated problem and we have a reasonable result. A better one may only be obtained with detailed calculations and serious computing. This case study attempted to show you that you have explored a topic with wide applicability. We hope (expect?) that it will whet your appetite to learn about quantum mechanics in more detail. It will be well worth the effort!

Exercises

1. The molecular structure of 1,3-butadiene was shown earlier in the chapter. Use the same average bond length proposed for the cyanine dyes (0.140 nm) to calculate, using the particle in a one-dimensional box model, the lowest energy transition for this molecule. Compare the result to the experimental wavelength value of 207 nm.
2. Use the valence bond method of assigning hybrid orbitals around each carbon atom to consider the bond angles in the cyclopropane molecule shown below. Would you assume that this is a stable, non-reactive molecule? Why or why not?

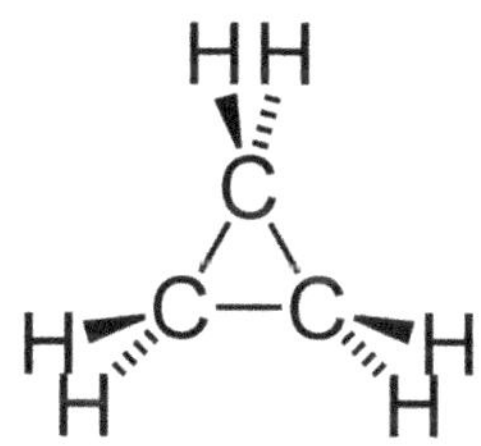

CHAPTER S1

Classical Waves

Because so much of what we will study relies on the assumption that an electron behaves like a wave, we thought it best to present some of the mathematical tools typically employed in discussions of ordinary or classical waves, such as those one would see in a vibrating guitar string. A plucked string is shown in Figure S1.1, and if you were to plot the displacement at the point $x = 0.5$ throughout time you would see an oscillating pattern, as shown in Figure S1.2, which is why the equation describing the plucked string is one of those generically called the "*wave equation*".

The equation which governs the displacement at point x and time t is given by:

$$\frac{\delta^2}{\delta x^2}F(x,t) = k\frac{\delta^2}{\delta t^2}F(x,t) \tag{S1.1}$$

This equation is derived from a discussion of forces acting on the guitar string in the form of tension, elasticity and other properties of guitar strings. These forces are usually described in

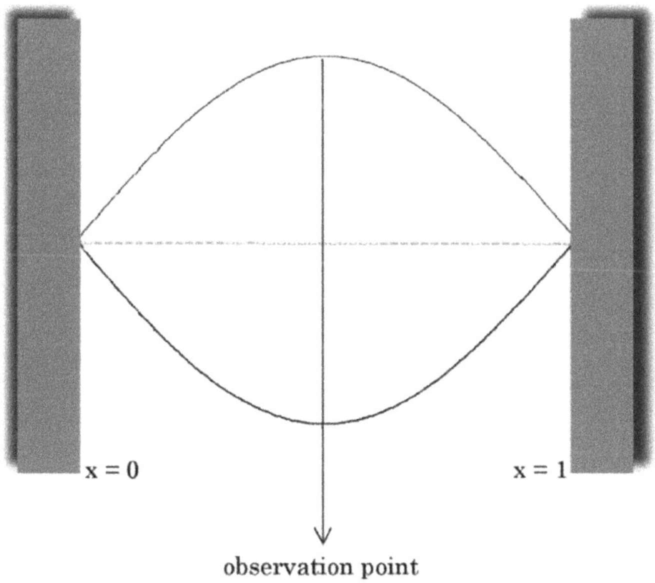

Figure S1.1. Three positions of an undamped string plucked at its midpoint.

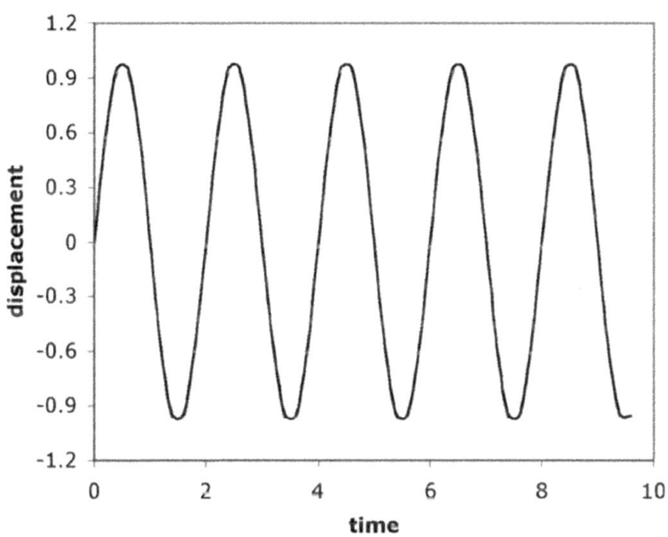

Figure S1.2. The temporal oscillations at $x = 0.5$ for the vibrating string in Figure S1.1.

terms of vectors acting on a point on the string. Since the equation involves an unknown function and a relationship between its partial derivatives, it is an example of a *partial differential equation.* We are going to ignore the model that gives us this equation because our intent here is not to examine how such a model is constructed, but rather to see where it leads in the solution of the problem. But there are three important details. First of all, the constant k is one of those numbers that is going to depend on the particulars of the string in question. Is it nylon or steel? Is it tight or loose? And so forth. Second, k is always positive. Third, this equation doesn't take damping into account, so when we finally get to a solution of it, the string will seem to continue to oscillate forever. Real strings won't do that.

What exactly constitutes a "problem" and what does it mean to find "the solution"? In this case, the *problem* is to describe the range of possible behaviors of a vibrating string, including the most observable of its behaviors, namely the pitch. Mathematically, this amounts to finding which functions could possibly describe the evolution of the position of the string over time, and what frequencies occur in the wave pictured above. Since any such function presumably satisfies the equation we wrote down, in terms of its second derivative in the x variable and its second derivative in the t variable, *a solution* means a function of x and t that satisfies that equation. To find *the solution* means to describe all such functions.

Before tackling the equation above, there is one more ingredient that is absolutely necessary to making a definitive statement about the nature of the vibrations. That is the *length of the string.* A guitar string is firmly secured in two places, let us say at $x = 0$ and at $x = 1$, just as is indicated in the drawing above. Mathematically, this means that the function $F(x, t)$ has to be equal to zero in certain places, namely:

$$F(0, t) = F(1, t) = 0 \tag{S1.2}$$

This is what is called a "boundary condition" for a partial differential equation. Such conditions are often as important in determining the nature of the solution as the actual equation itself. Here is the assumption behind what happens next: *If one can find one kind of solution to that equation, one can figure out how to get all of them.* So the general strategy is to find one solution, or one family of solutions, then use them to get a very large set of solutions. Then, after the fact, go back and prove that you did, indeed, find all of the possible solutions.

This is called "separation of variables". One assumes that the function $F(x,t)$ is really the product of two functions: $f(x)$ and $g(t)$. Not only that, but we will assume that f obeys the boundary conditions above. Now, Equation (S1.1) above becomes:

$$\frac{\delta^2}{\delta x^2} f(x) g = k \frac{\delta^2}{\delta t^2} f g(t) \tag{S1.3}$$

$$f(0) = f(1) = 0$$

where the derivative in x only affects the function f and the derivative in t only affects the function g. Using some simple algebra gives you:

$$\frac{\frac{\delta^2}{\delta x^2} f}{f} = k \frac{\frac{\delta^2}{\delta t^2} g}{g} \tag{S1.4}$$

The left-hand side of this equation depends only on x and the right-hand side depends only on t. So both sides must be equal to the same constant. Now you have two ordinary differential equations in one unknown each, and you have an unidentified parameter, namely the constant that both sides of the equation must equal. Let's call this constant C. Now we have two separate equations, each in only one variable. The first one is:

$$\frac{\frac{\delta^2}{\delta x^2} f}{f} = C \tag{S1.5}$$

$$f(0) = f(1) = 0$$

And the second is:

$$k\frac{\frac{\delta^2}{\delta t^2}g}{g} = C \tag{S1.6}$$

Of course, we could rewrite these equations by multiplying both sides by either f or g, as appropriate. Then the first would look like:

$$\frac{\delta^2}{\delta x^2}f = Cf \tag{S1.7}$$

An equation of this sort, which has some expression in f on the left similar to this one and a constant times f on the right, is called an *eigenvalue* problem. C is the "eigenvalue" in this problem. The function f is called the "eigenfunction".

Let's look at that first equation. If C is negative then it looks just like the equation for a spring that comes from Hooke's law. That is, there are two functions that solve this equation which are familiar to everyone:

$$f_1(x) = \sin\sqrt{|C|}x$$

and

$$f_2(x) = \cos\sqrt{|C|}x \tag{S1.8}$$

Furthermore, all combinations of the form $af_1 + bf_2$ will also solve the equation in x. However, to get the boundary conditions we have to make sure that $f_1(x)$ and $f_2(x)$ assume a value of zero at $x = 0$ and $x = 1$. This puts restrictions on what C could be. If you think about it, C must now be a negative integer multiple of π. Furthermore, the cosine solutions don't work at all.

But what if C is positive? Well, then the solution looks very different. It is easy to check that the two functions

$$f_1(x) = e^{(\sqrt{C}x)}$$

and

$$f_2(x) = e^{(-\sqrt{C}x)} \tag{S1.9}$$

do the job quite nicely, except that neither of them satisfies the boundary conditions. So, we reject these solutions as possibilities.

Consider this concept of rejecting solutions that aren't as "nice" as we might like. Solving a problem of this sort requires making a number of somewhat arbitrary and not wholly justified decisions, proceeding to see what happens, and finally checking to see if your guesses were good. Historically, this is what seems to have happened. In very applied contexts, such as engineering, it is still the way to proceed. Because this problem has been very well studied, however, it is possible to use mathematics to justify some of these decisions upfront. For example, the entire mathematical field called *harmonic analysis* stems from the proof of the observation that the sines and cosines we found for our solution when C was negative, taken together, are big enough to build all the other solutions by taking infinite linear combinations of them.

Let's summarize where we are. By a series of assumptions and calculations we have narrowed down our acceptable solutions to those of the form:

$$f_1(x) = \sin \sqrt{|C|}x \tag{S1.10}$$

where C must be a negative integer multiple of π. Now we must turn our attention to the equation in the variable t:

$$k\frac{\frac{\delta^2}{\delta t^2}g}{g} = C \tag{S1.11}$$

we can rewrite as

$$\frac{\delta^2}{\delta t^2}g = \frac{C}{k}g \tag{S1.12}$$

and as before we get solutions of the form:

$$g_1(t) = \sin\left(\sqrt{\frac{|C|}{k}}t\right)$$

and

$$g_2(t) = \cos\left(\sqrt{\frac{|C|}{k}}t\right) \tag{S1.13}$$

The boundary conditions are in the x variable and do not affect g, so we do not need to eliminate the cosine solutions this time. Now we have a large family of solutions to this equation, given by:

$$F_{n,1}(x,t) = \sin(n\pi x)\sin\left(\frac{n\pi}{\sqrt{k}}t\right)$$

and

$$F_{n,2}(x,t) = \sin(n\pi x)\cos\left(\frac{n\pi}{\sqrt{k}}t\right) \tag{S1.14}$$

Out of these infinitely many solutions we can construct a generic one that looks like:

$$\sum_n a_n F_{n,1}(x,t) + b_n F_{m.2}(x,t) \tag{S1.15}$$

which can be expanded as:

$$F(x,t) = \sum_n a_n \sin(n\pi x)\sin\left(\frac{n\pi}{\sqrt{k}}t\right) + b_n \sin(n\pi x)\cos\left(\frac{n\pi}{\sqrt{k}}t\right) \tag{S1.16}$$

Now if you are not totally convinced that anything of this form is in fact a solution to the original wave equation, you should plug it in and see for yourself. Once you convince yourself of that, the only remaining question is whether such a scheme gives every possible solution. There are two ways to think about this question. Briefly, one way uses physics, and the other uses math.

First let's talk about the physics. If you believe that the wave equation above is a good model for a vibrating string, and if you believe in physics, then you should also believe that, given an initial state for the string, there will be only one possible behavior stemming from it. By "initial state" we mean a description of how the string is deformed at $t = 0$. If you pluck a guitar string in a certain way and let it go, it should obey whatever physical laws control it and *always do exactly the same thing*. So, if you can match any initial condition, then you have described all possible solutions of the problem. Setting $t = 0$ gives:

$$F(x,0) = \sum_{n>0}^{n} b_n \sin(n\pi x) \tag{S1.17}$$

The theory of the Fourier series tells us that we can find all of the different b_n needed to match any initial condition $F(x,0)$. This approach doesn't tell us anything about any a_n. The physical explanation for that is that we could also have given our string an initial velocity distribution, $F_t(x,0)$. This can be achieved with a real guitar by musical techniques such as "hammering on" or "pulling off", and must be taken into account. To do this we have to compute:

$$F(x,0) = \sum_{n>0}^{n} a_n \frac{n\pi}{\sqrt{k}} \sin(n\pi x) \tag{S1.18}$$

Now we can match two sets of initial conditions at once and get a unique answer using the Fourier series. This is a physical approach.

A more mathematical approach would invoke a theorem of differential equations, which says that a second-order partial differential equation such as the one we have here, with two initial conditions of the form we just used, must have a unique solution. The branch of mathematics you would have to study to learn this theorem is called *partial differential equations*. Mathematically, this type of theorem makes the claim for our *model of nature* that the physical explanation is attempting to make for *nature herself.* Then we would invoke the theory of the Fourier series to tell us that the sines and cosines are good enough to solve the problem.

Exercises

1. Show that the function described by Equation (S1.8) satisfies the boundary condition $f_1(0) = 0$. Show that the function $f_2(0)$ in Equation (S1.9) does not satisfy this condition.

CHAPTER S2

The Spherical Harmonics

The spherical harmonics are really the families of functions of the Euler angle variables (θ, ϕ) which are invariant under rotation. The orthogonal group is not commutative, so its action on these variables doesn't amount to simply adding angles independently to the Euler angle variables. Rather, there is some complex interaction between rotations in θ and ϕ. A simple approach is to view every function on the sphere as the restriction of a function in three-dimensional space, where the action of the orthogonal group is given by a matrix by which one multiplies the vector (x, y, z).

A function, $f(x, y, z)$ becomes $f(Ax + By + Cz, Dx + Ey + Fz, Gx + Hy + Iz)$ where the matrix $\rho(g)$ is given by:

$$\begin{pmatrix} A & B & C \\ D & E & F \\ G & H & I \end{pmatrix} \tag{S2.1}$$

This function, when restricted to values of (x, y, z) where $x^2 + y^2 + z^2 = 1$, gives a function of the points on the sphere. Of course, we will have to be careful because more than one function might be the same when restricted to just the sphere. So we will have to worry about that when we make our count of how many there are in each invariant subspace.

Suppose we look at a particular function:

$$f(x, y, x) = x^3 \tag{S2.2}$$

Then, if we transform the function according to the matrix above,

$$\rho(g)f(x, y, x) = (Ax + By + Cz)^3 \tag{S2.3}$$

What aspects of f are preserved by such a transformation? Multiplying the terms out gives us a major hint. Every term of the transformed polynomial has the same degree as the original, namely 3. So, if we took as a subspace of our huge vector space of functions those polynomials which are *homogeneous of degree n*, which means that every term has exponents which add up to n, then we would have a subspace that is invariant under our group action. Now, we have to be careful because some of them are really the same function when confined to the sphere. For example,

$$f(x, y, z) = x^2 + y^2 + z^2 \tag{S2.4}$$

is really a constant function on the sphere, so its degree goes down by 2. Because we believe in the mathematical phenomenon known as the Taylor Series, we know that these polynomials are enough to generate all functions on the sphere.

It's worth noting the weirdness of this example. Normally if you take a second derivative of a polynomial of a certain degree, that degree goes down by 2. In other words, this trick of using polynomials won't work for the regular Laplace operator, so why should it work for the spherical one (mathematicians call this function Δ, this is the function we chemists call ∇^2)? The key lies in the finite dimensional hypothesis in Schur's lemma. The lemma itself is in abstract algebra textbooks; we only need to know that for the lemma to work, the group G and the operator Δ or ∇^2 must commute on this finite dimensional space. Recall that the operator is:

$$\frac{\delta^2}{\delta x^2} + \frac{\delta^2}{\delta y^2} + \frac{\delta^2}{\delta z^2} \tag{S2.5}$$

and does not preserve the space of polynomials of any particular degree, so we can't use Schur's lemma on it. In other words, although the group breaks the space of polynomials of degree *less than or equal to* x into invariant subspaces, the Laplacian doesn't cooperate by respecting those subspaces, at least not superficially. In fact, if it is going to be true that the Laplace operator has an eigenfunction that is a polynomial, then the eigenvalue must be *zero*. Do such functions exist? Yes they do, and they are called the *harmonic polynomials*. Here are some examples:

$$\begin{gathered} x, y, z \\ x - y^2, y - z^2, xy, yz \\ 3x^2y - y^3 \\ x^3y - xy^3 \end{gathered} \tag{S2.6}$$

and so on. Since each of these is an eigenfunction of Δ, if we use any one of them to make more by changing variables under the rotation group, we will construct many new eigenfunctions in this way. Now we have made spaces which satisfy the hypotheses of Schur's lemma, although the eigenvalue corresponding to the operator Δ is always zero.

As a matter of fact, if we look at what happens to those polynomials when we substitute Euler angles and set the radius to 1, we find something quite different. The polynomial z, for example, becomes:

$$f(\theta, \phi) = \cos(\theta) \tag{S2.7}$$

If we apply the spherical Laplacian, Δ_S, we get:

$$\Delta_S f = -2\cos(\theta) \tag{S2.8}$$

If we pass to polynomials of degree 2, we have to be careful. They do not always look homogeneous when we finish setting $x^2 + y^2 + z^2 = 1$. Let us look at z^2, for example, which is not harmonic and so should not be an eigenfunction of Δ. Passing to spherical coordinates, and setting $r = 1$, this gives the function:

$$g(\theta, \phi) = \cos^2(\theta) \tag{S2.9}$$

You can check that:

$$\Delta_S g = 2 - 6\cos^2(\theta) \tag{S2.10}$$

Clearly if we try a new function:

$$G(\theta, \phi) = \cos^2(\phi) - \frac{1}{3} \tag{S2.11}$$

Then G is an eigenfunction of Δ_S with eigenvalue -6. Now, if we were to create a bunch of new functions by applying our rotation group to this one, we would indeed create the irreducible representation described by Schur's lemma. Does this correspond to a harmonic function in x, y, z? Well, you can check that

$$z^2 = \frac{1}{3}(x^2 + y^2 + z^2) \tag{S2.12}$$

is in fact harmonic, and restricts to

$$z^2 - \frac{1}{3} \tag{S2.13}$$

on the unit sphere. It is easy to construct a harmonic polynomial for any given degree, but hard to write down a general recipe for getting one of an arbitrary degree.

In fact, eigenfunctions for the spherical Laplacian can be written down explicitly in terms of *Legendre polynomials*. The Legendre polynomials are defined in terms of two indices as follows:

$$P_n^M(w) = (1 - x^2)\frac{d^m P_n(x)}{dx^m} \tag{S2.14}$$

where

$$P_n(x) = \frac{1}{2^n n!}\frac{d^n}{dx^n}((x^2 - 1)^n) \tag{S2.15}$$

Armed with all of this notation, the function

$$e^{im\phi}P_n^m(\cos(\theta)) \tag{S2.16}$$

is an eigenfunction of Δ_S with eigenvalue $n(n+1)$.

So, like many math problems, this is one you can "solve" once you know the answer. And, like a lot of math problems, the modern approach (which in this case uses group theory and the fancy algebra of Schur's lemma) harkens back to famous families of "special functions". You can also do a fancy counting argument to find the *dimension* of each space of eigenfunctions, and you will find that the dimension of the space of functions whose eigenvalue is $-n(n+1)$ is, in fact, $2n+1$.

Without even solving the equation in the radial variable, just knowing the spherical harmonics gives us a lot of information about the structure of the atom. The probability distribution of the electron is usually pictured as a cloud, where the density of the cloud denotes greater probability. Nodes of the spherical functions are choices of θ and ϕ where the solution is zero. For example:

$$f(\theta, \phi) = \cos(\theta) \tag{S2.17}$$

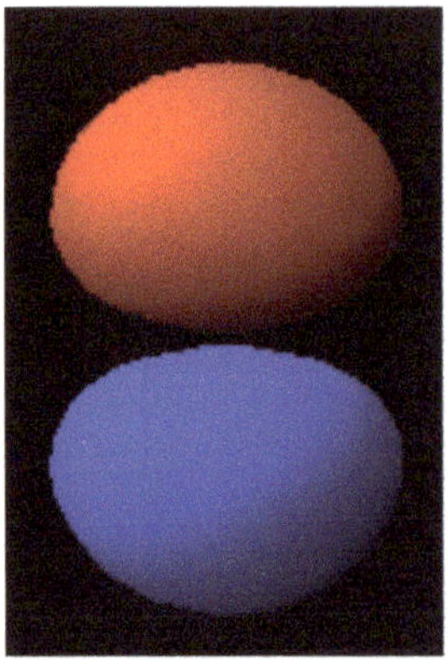

Figure S2.1. Graphical representation of Equation (S2.17). This is also the representation of a stylized $2p$ orbital.

has a node at $\theta = \pi/2$, which is the entire xy plane. The electron cloud for this solution looks like that shown in Figure S2.1 and represents one of the orbitals portrayed in any chemistry or physics textbook.

Solutions to Exercises

Chapter 2

2.1 Equation (2.3) provides the relationship between wavelength and frequency, $\nu = c/\lambda$ where c is the speed of light or $\sim 3 \times 10^8 \mathrm{m\,s^{-1}}$.

$$\nu_{650\,\mathrm{nm}} = \frac{3 \times 10^8\,\mathrm{m/s}}{650 \times 10^{-9}\,\mathrm{m}} = 4.6 \times 10^{14}\,\mathrm{s^{-1}}$$

$$\nu_{230\,\mathrm{nm}} = \frac{3 \times 10^8\,\mathrm{m/s}}{230 \times 10^{-9}\,\mathrm{m}} = 1.3 \times 10^{15}\,\mathrm{s^{-1}}$$

Using Figure 2.4 for reference, 650 nm is in the red region of visible light and 230 nm is in the ultraviolet region of the spectrum.

2.2 Equation (2.4) states that the radiative density is given by

$$\rho(\nu, T)d\nu = \frac{8\pi kT}{c^3}\nu^2 d\nu$$

$$\rho(1/\lambda, T) = \frac{8\pi kT}{c^3}\frac{c^2}{\lambda^2} = \frac{8\pi(1.38 \times 10^{-23}\,\mathrm{J/K})(300\,\mathrm{K})}{3 \times 10^8\,\mathrm{m/s}\,(\lambda^2)}$$

$$= \frac{3.46 \times 10^{-28}}{\lambda^2}$$

$$\rho_{100\,\mathrm{nm}} = 3.47 \times 10^{-14}\,\mathrm{J\,m^{-3}}$$

$$\rho_{1000\,\mathrm{nm}} = 3.47 \times 10^{-16}\,\mathrm{J\,m^{-3}}$$

We observe that as the wavelength decreases, the radiative density is increasing dramatically and it does not reach a maximum. Figure 2.5 includes a blackbody curve for 1000 K and we see that there is a wavelength of maximum intensity.

2.3 Equation (2.6) is Einstein's photoelectric effect description:

$$\frac{1}{2}mv^2 = h\nu - \phi$$

where $h\nu$ is the incident photon. If $h\nu - \phi \geq \phi$, the electron is ejected.

Note that $1\,\text{eV}$ is $1.602 \times 10^{-19}\,\text{J}$, so $4.7\,\text{eV} = \phi = 7.52 \times 10^{-19}\,\text{J}$.

If the wavelength is 300 nm, then the energy of the photon is $E = \frac{hc}{\lambda} = 6.62 \times 10^{-19}\,\text{J}$, so this is insufficient to eject the electron.

If the wavelength is 132 nm, then the energy of the photon is $E = \frac{hc}{\lambda} = 15.1 \times 10^{-19}\,\text{J}$, so this is sufficient to eject the electron.

2.4 The de Broglie equation, Equation (2.7), provides the momentum:

$$\rho = \frac{h}{\lambda}$$

At 700 nm,

$$\rho = \frac{6.62 \times 10^{-35}\,\text{J s}}{700 \times 10^{-9}\,\text{m}} = 9.46 \times 10^{-28}\,\text{J s m}^{-1}$$

At 350 nm,

$$\rho = \frac{6.62 \times 10^{-35}\,\text{J s}}{350 \times 10^{-9}\,\text{m}} = 18.9 \times 10^{-28}\,\text{J s m}^{-1}$$

Chapter 3

3.1 Balmer's expression is in Equation (3.1a):

$$\frac{1}{\lambda} = 109,680\left(\frac{1}{4} - \frac{1}{n^2}\right)\ \mathrm{cm}^{-1}, \quad n = 3, 4, 5, \ldots$$

We wish to extend the calculation to $n = 6$:

$$\frac{1}{\lambda} = 109{,}680\left(\frac{1}{4} - \frac{1}{6^2}\right)\mathrm{cm}^{-1}$$

$$= 109{,}680(0.22)\,\mathrm{Hz} = 24372.3\,\mathrm{cm}^{-1}$$

$$\lambda = 4.10 \times 10^{-5}\,\mathrm{cm} = 410\,\mathrm{nm}$$

This wavelength was just outside of the region that Balmer's apparatus (his spectrometer) could detect.

3.2 The text indicates that the expression for the hydrogen atom Rydberg constant is

$$\frac{1}{2}\frac{Z^2e^2}{a_0} = R_H$$

$Z = 1$ when we focus on hydrogen, so in general, $R = R_H \cdot Z^2$.

3.3 We find the general expression for the change in energy is

$$\Delta E = E_f - E_i = \frac{1}{2}\frac{Z^2e^2}{a_0}\left(\frac{1}{n_f^2} - \frac{1}{n_i^2}\right)$$

So, Z^2 will be 4 when we have the He^+ ion ($Z = 2$), $R_{He+} = 4R_H$, and the change in energy will be 4× that for one hydrogen atom. $\Delta E_H = 4.08 \times 10^{-19}$. So for He^+, $\Delta E = 1.63 \times 10^{-18}$ J.

$$\Delta E = \frac{hc}{\lambda}$$

$$\lambda = \frac{hc}{\Delta E} = \frac{(6.62 \times 10^{-35}\,\mathrm{J\,s})(3 \times 10^{8}\,\mathrm{m/s})}{1.63 \times 10^{-18}\,\mathrm{J}}$$
$$= 121 \times 10^{-9}\ \mathrm{m} = 121\,\mathrm{nm}$$

This wavelength cannot be observed in the visible region of the spectrum. It will be found in the high-energy vacuum ultraviolet region.

3.4 The absolute minimum wavelength is the size of the required uncertainty, 0.001 nm. The wavelength should actually be shorter than that size, but we'll use it for the calculation. The Uncertainty Principle states that

$$\Delta x \Delta p = (0.001\,\mathrm{nm})(\Delta p) \geq \frac{\hbar}{2}$$
$$(\Delta p) \geq \frac{6.62 \times 10^{-34}\,\mathrm{J\,s}}{4\pi(0.001 \times 10^{-9}\,\mathrm{m})}$$
$$= 5.27 \times 10^{-23}\,\mathrm{J\,s\,m^{-1}} \quad \text{or} \quad 5.27 \times 10^{-23}\,\mathrm{kg\,m\,s^{-1}}$$

Chapter 4

4.1 The commutation relationship is Equation (4.1). If two operators commute, they can be determined exactly and simultaneously. But the relationship shows that the position and momentum operators are not commutative.

$$\hat{x}\hat{p} - \hat{p}\hat{x} = i\hbar$$

The Heisenberg Uncertainty Principle, which states that some selected properties cannot be measured or known exactly and simultaneously, is

$$\Delta x \Delta p \geq \frac{\hbar}{2}$$

These two equations are relating exactly the same principle.

4.2 The wavefunction is given by

$$\Psi = Ce^{i(px-Et/\hbar)}$$

The derivative we seek is

$$\begin{aligned}\frac{\delta\Psi}{\delta x} &= C\frac{\delta}{\delta x}e^{i(px-Et/\hbar)} = C\frac{\delta}{\delta x}(e^{i(px}e^{-iEt/\hbar})\\ &= Ce^{-iEt/\hbar}\frac{\delta}{\delta x}(e^{ipx}) = Ce^{-iEt/\hbar}e^{ipx}\frac{\delta}{\delta x}(ipx)\\ &= ipCe^{-iEt/\hbar}e^{ipx} = ip\Psi\end{aligned}$$

4.3 The wavefunction is given by

$$\Psi = Ce^{i(px-Et/\hbar)}$$

The derivative we seek is

$$\begin{aligned}\frac{\delta}{\delta l}\Psi &= C\frac{\delta}{\delta l}(e^{i(px-Et/\hbar)}) = C\frac{\delta}{\delta t}(e^{i(px}e^{-iEt/\hbar})\\ &= Ce^{ipx}\frac{\delta}{\delta t}(e^{-iEt/\hbar}) = Ce^{ipx}e^{-iEt/\hbar}\frac{\delta}{\delta t}(-iEt/\hbar)\\ &= Ce^{ipx}e^{-iEt/\hbar}\frac{\delta}{\delta t}(-iEt/\hbar) = -i\frac{E}{\hbar}Ce^{i(px-Et/\hbar)} = -i\frac{E}{\hbar}\Psi\end{aligned}$$

4.4 Equation (4.16) provides

$$\hat{p}^2 \rightarrow (i\hbar)\left(\frac{\delta}{\delta x}\right)\cdot(i\hbar)\left(\frac{\delta}{\delta x}\right)$$

If we operate with $\hat{p}^2$ on some function Ψ, we get

$$\hat{p}\Psi = (-i\hbar)\left(\frac{\delta}{\delta x}\right)\Psi$$

Then operate a second time:

$$\begin{aligned}\hat{p}\left[(-i\hbar)\left(\frac{\delta}{\delta x}\right)\right]\Psi &= (-i\hbar)\left(\frac{\delta}{\delta x}\right)(-i\hbar)\left(\frac{\delta}{\delta x}\right)\Psi\\ &= \mathrm{i}^2\hbar^2\left(\frac{\delta^2}{\delta x^2}\right)\Psi\end{aligned}$$

Chapter 5

5.1 The wavefunction must be equal to zero when $x = L$ and the wavefunction is defined in the discussion around Equation (5.1):

$$\Psi(L) = Dsin(kL)$$

The function $\sin(x) = 0$ when $x = 0$ or some integer multiple of π. Therefore,

$$\Psi(L) = 0, \quad \text{if } kL = n\pi \rightarrow k = \frac{n\pi}{L}$$

From Equation (5.1),

$$k = \left(\frac{2mE}{\hbar^2}\right)^{1/2}$$

and

$$k^2 = \left(\frac{2mE}{\hbar^2}\right) = \frac{n^2\pi^2}{\hbar^2} \rightarrow E = \frac{n^2\pi^2\hbar^2}{2mL^2}$$

5.2 From Equation (5.2),

$$1 = \int \Psi * \Psi dx = D^2 \int_0^L \sin^2\left(\frac{n\pi x}{L}\right) dx$$

From a table of integrals,

$$\int \sin^2 ax\, dx = \frac{x}{2} - \frac{\sin(2ax)}{4a}$$

Evaluating between our limits of 0 to L,

$$1 = D^2 \left[\frac{L}{2} - \frac{\sin\left(\frac{2n\pi L}{L}\right)}{\frac{4n\pi}{L}} - \frac{0}{2} + \frac{\sin(0)}{\frac{4n\pi}{L}}\right]$$

$$= D^2 \left[\frac{L}{2} - \frac{\sin(2n\pi)}{4n\pi/L} - 0 - 0\right] = D^2\frac{L}{2} \rightarrow D = \left(\frac{2}{L}\right)^{1/2}$$

5.3 Equation (5.3) is

$$\Delta E = E_{n+1} - E_n = \frac{(2n+1)\pi^2\hbar^2}{2mL^2}$$

$$E_{n+1} = \frac{(n+1)^2\pi^2\hbar^2}{2mL^2}$$

$$E_n = \frac{n^2\pi^2\hbar^2}{2mL^2}$$

$$\Delta E = E_{n+1} - E_n = \frac{(n+1)^2\pi^2\hbar^2}{2mL^2} - \frac{n^2\pi^2\hbar^2}{2mL^2}$$

$$= \frac{(n^2+2n+1)\pi^2\hbar^2}{2mL^2} - \frac{n^2\pi^2\hbar^2}{2mL^2} = \frac{(2n+1)\,\pi^2\hbar^2}{2mL^2}$$

5.4 Since the ground state has $n_x = n_y = n_z = 1$, the total ground state energy is the sum of three terms where $n^2 = 1$.

$$E_{GS} = \frac{3\pi^2\hbar^2}{2mL^2}$$

The first excited state has $n = 2$ in one direction and $n = 1$ in the other two directions, so the total energy in this state will be

$$\frac{(2^2+1+1)\pi^2\hbar^2}{2mL^2}$$

The second excited state has $n = 2$ in two directions and $n = 1$ in the other direction:

$$E - \frac{(2^2+2^2+1)\pi^2\hbar^2}{2mL^2} = \frac{9\pi^2\hbar^2}{2mL^2}$$

$$\Delta E = \frac{(9-3)\pi^2\hbar^2}{2mL^2}$$

5.5 Refer to Equation (5.4), where $L_x \neq L_y \neq L_z$, but $n_x = n_y = n_z = 1$:

$$E = \frac{\pi^2\hbar^2}{2mL_x^2} + \frac{\pi^2\hbar^2}{2mL_y^2} + \frac{\pi^2\hbar^2}{2mL_z^2}$$

Chapter 6

6.1 Ground state: $n = 1$; $\ell = 0$; $m_l = 0$; $m_s = \pm 1$

1st excited state: $n = 2$; $\ell = 0, 1$; $m_l = -1, 0, 1$; $m_s = \pm 1$
2nd excited state: $n = 3$; $\ell = 0, 1, 2$; $m_l = -2, -1, 0, 1, 2$; $m_s = \pm 1$

6.2 The orbitals shown in Figure 6.3 are hydrogen s orbitals. The nodes for s orbitals are found in the radial wavefunctions and there are $n - 1$ nodes. If we count the number of nodes, that will reveal the identity of the curves. The $2s$ orbital should have one node.

6.3 The spherical coordinates are defined in Equation (6.2) and squaring each yields

$$\begin{aligned} x^2 &= r^2 \sin^2\theta \cos^2\phi \\ y^2 &= r^2 \sin^2\theta \sin^2\phi \\ z^2 &= r^2 \cos^2\theta \end{aligned}$$

Let's do partial sums ($x^2 + y^2$, then add z^2 to this sum):

$$\begin{aligned} x^2 + y^2 &= r^2 \sin^2\theta(\cos^2\phi + \sin^2\phi) = r^2 \sin^2\theta \\ x^2 + y^2 + z^2 &= r^2 \sin^2\theta \cos^2\theta = r^2 \\ r &= \sqrt{x^2 + y^2 + z^2} \end{aligned}$$

6.4 The average position is important because the energy of the electron depends only on the distance from the nucleus, not any directionality. We cannot know the exact position of the electron, but if we know its average position, we can calculate its average energy in that orbital.

6.5 The energy of the electron in a one-electron atom or ion depends only on the principal quantum number, n. Therefore, the answer is that the energies are the same for all of the set of quantum numbers where $n = 2$.

6.6 As shown in the text, the most probable radius is found by taking the derivative of the radial probability and setting it equal to zero, the standard calculus method for finding maxima. For the hydrogen atom, this yields a value of $r_{mp} = a_0$, the Bohr radius. This result is what Bohr also predicted without wave mechanics.

6.7 The node occurs when the wavefunction is equal to zero. Since this is a one-electron atom, we can find the node in the radial wavefunction.

$$R(r) = \frac{1}{2\sqrt{2}}\left(\frac{1}{a_0}\right)^{3/2}\left(2 - \frac{r}{a_0}\right)e^{-r/2a_0} = 0$$

This will be zero when

$$\left(2 - \frac{r}{a_0}\right) = 0 \rightarrow r = 2a_0$$

6.8 The $2p$ radial wavefunction is

$$R(r) = \frac{1}{2\sqrt{6}}\left(\frac{1}{a_0}\right)^{3/2}\left(\frac{r}{a_0}\right)e^{-r/2a_0}$$

This will only be zero at $r = 0$. The $2p_z$ wavefunction is

$$Y_{l=1} = \left(\frac{3}{4\pi}\right)^{1/2}\cos(\theta)$$

This will have a node and will equal zero when

$$\cos(\theta) = 0 \quad \text{or} \quad \theta = \frac{\pi}{2}$$

See also Figure 6.4 for a graphical response to this exercise.

Chapter 8

1. Neon has eight outer ($n = 2$) electrons.
   Na^+ is isoelectronic with neon.
   Boron has three outer ($n = 2$) electrons.
   Oxygen has six outer electrons.

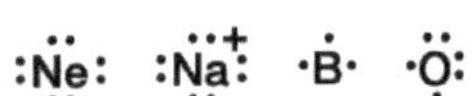

2. All of these atoms/ions have 14 electrons (and have triple bonds):

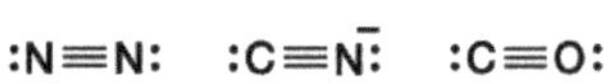

3. The important point in this particular exercise is that if the number of electrons in the molecule is insufficient to have all of the atoms satisfy the octet rule, the octet rule for the more electronegative atom will be satisfied first. In the first three molecules shown below, all atoms have filled octets. The deficiency described at the start of this solution occurs for the second set of three molecules/ions.

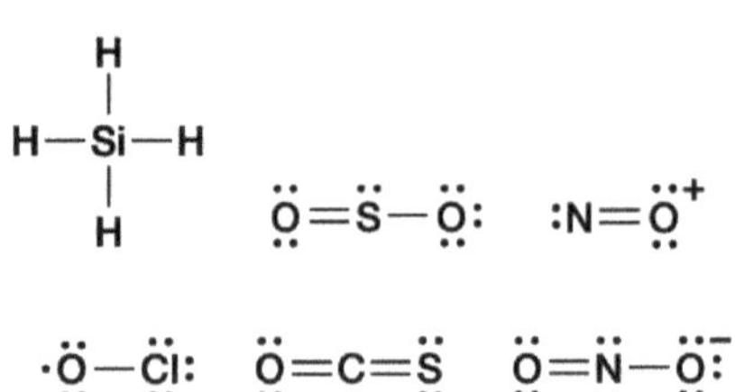

4. The sulfur atom has electrons in higher orbitals than $n = 2$, and can have an "expanded octet", meaning more than eight electrons associated with the atom.

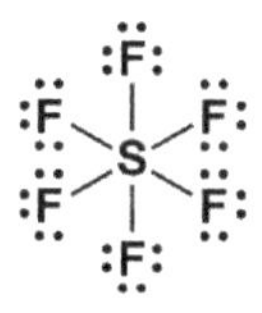

5. The solution to these exercises requires the use of multiple bonds to maintain the number of available electrons. N_2O_4 has 34 available electrons, C_2H_4 has 12 available electrons and SO_4^{2-} has 34 available electrons (including the charge of 2−).

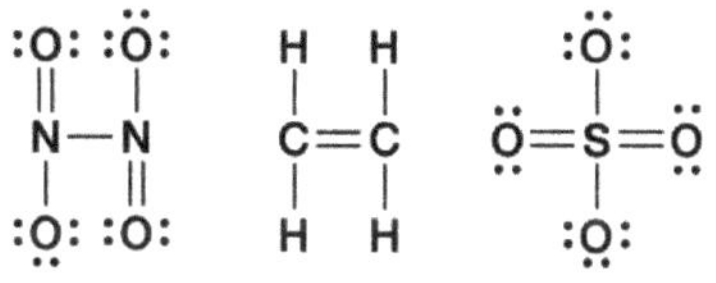

Chapter 9

1. These molecules will all have geometries distorted from the ideal geometry predicted from the number of "bonding electrons" due to non-bonding electrons. H_2O has eight total electrons available, but is only bonded to two atoms. A tetrahedral geometry is the ideal, but the non-bonding electrons cause a distortion to a bent geometry. In XeF_4, the Xe atoms can have an expanded octet, but counting electrons (36 total) there are two pairs of non-bonding electrons. If we place the non-bonding electrons 180° apart, the molecule will have a square planar geometry (distorted from the expected octahedral geometry for 12 electrons). Finally, ClF_3 is a unique case. The central atom is Cl because the fluorine atom is more electronegative. There are 26 total

electrons. Like the H_2O molecule, the observed geometry will be distorted from tetrahedral and will be pyramidal.

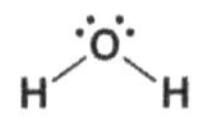 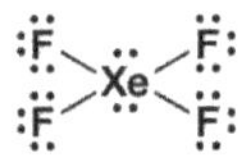 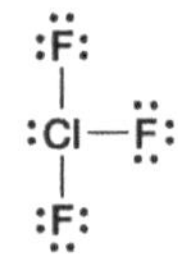

2. The geometries and structures of the first three are straightforward with only a single twist. CO_2 and CS_2 are closely related. Both have 16 total electrons (no need for an expanded octet on the sulfur atom) and linear geometries. SO_2 has 18 total electrons. The central sulfur atom has a non-bonding pair of electrons so that the geometry is distorted from the ideal trigonal and the molecule is bent. H_2CO has 12 total electrons. Since the carbon atom is the central atom, a double bond is needed and is located between the carbon and the oxygen atoms.

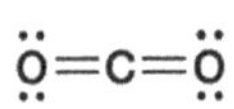 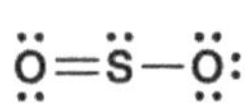 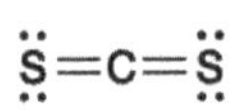 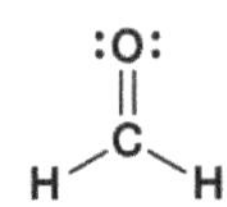

3. The ideal geometry of AsF_3 is tetrahedral, but the molecule has 26 total electrons and the central As atom has a non-bonding pair of electrons. Note that As is the same group of the Periodic Table as nitrogen. The observed geometry will be pyramidal (distorted terahedral). NNO has a total of 16 electrons and a triple bond is needed. It could be between a pair of nitrogen atoms or between nitrogen and oxygen. These could be resonance structures. We show one of the linear geometries below.

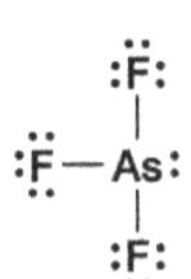 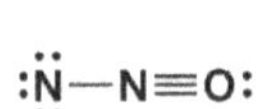 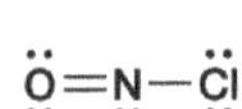

Chapter 10

1. Using Figure 10.12 and the observation that Be is a light atom with four electrons, Be_2 will have eight total electrons and the electron configuration $\sigma_{1s}^2\sigma_{1s}^{*2}\sigma_{2s}^2\sigma_{2s}^{*2}$ with a bond order of $(4-4)/2 = 0$. Be_2^+ has one fewer electrons with the electron configuration $\sigma_{1s}^2\sigma_{1s}^{*2}\sigma_{2s}^2\sigma_{2s}^{*}$ and a bond order of $(4-3)/2 = 1/2$. Finally, Be_2^- has one additional electron or the configuration $\sigma_{1s}^2\sigma_{1s}^{*2}\sigma_{2s}^2\sigma_{2s}^{*2}\pi_{2p}$. The bond order is $(5-4)/2 = 1/2$. However, you should note that this bond involves a π orbital. The half-order bond is a single-electron π bond.

2. If a molecule has at least one "heavy atom", as in CO, we will treat it as a heavy atom molecule. The electron configurations for the ions in this exercise are

CO^+	13 electrons	$\sigma_{1s}^2\sigma_{1s}^{*2}\sigma_{2s}^2\sigma_{2s}^{*2}\sigma_{2p_z}^2\pi_{2p}^3$
CO^-	15 electrons	$\sigma_{1s}^2\sigma_{1s}^{*2}\sigma_{2s}^2\sigma_{2s}^{*2}\sigma_{2p_z}^2\pi_{2p}^4\pi_{2p}^*$
CN^+	12 electrons	$\sigma_{1s}^2\sigma_{1s}^{*2}\sigma_{2s}^2\sigma_{2s}^{*2}\pi_{2p}^4$
CN^-	14 electrons	$\sigma_{1s}^2\sigma_{1s}^{*2}\sigma_{2s}^2\sigma_{2s}^{*2}\pi_{2p}^4\sigma_{2p_z}^2$
CO	14 electrons	$\sigma_{1s}^2\sigma_{1s}^{*2}\sigma_{2s}^2\sigma_{2s}^{*2}\sigma_{2p_z}^2\pi_{2p}^4$
CN	13 electrons	$\sigma_{1s}^2\sigma_{1s}^{*2}\sigma_{2s}^2\sigma_{2s}^{*2}\pi_{2p}^4\sigma_{2p_z}$

   CO and CN^- are isoelectronic; even though the orbitals fill in a different order, the same MOs are filled. CO^+ and CN have the same number of electrons, but different electron configurations.

3. O_2 is one of the heavy molecules. Its 16 electrons would have the $\sigma_{1s}^2\sigma_{1s}^{*2}\sigma_{2s}^2\sigma_{2s}^{*2}\sigma_{2p_z}^2\pi_{2p}^4\pi_{2p}^{*2}$ electron configuration and a bond order of $(10-6)/2 = 2$. If we remove an electron to make the O_2^+ ion, there would be one fewer π antibonding electron and the bond order would increase to $(10-5)/2 = 2.5$. If we add an electron to make the O_2^- ion, there would be one more π antibonding electron (compared with O_2) and the bond order

would decrease $(10-7)/2 = 1.5$. For the nitrogen molecule, we begin with a bond order of $(10-4)/2 = 3$ because the electron configuration is $\sigma_{1s}^2\sigma_{1s}^{*2}\sigma_{2s}^2\sigma_{2s}^{*2}\pi_{2p}^4\sigma_{2p_z}^2$. Removing an electron and creating N_2^+ will decrease the bond order since a bonding electron is ionized. The bond order will be $(9-4)/2 = 2.5$. Creating the N_2^- ion will have the same effect on the bond order because the added electron will be placed in the π_{2p}^* orbital and the bond order calculation yields $(10-5)/2 = 2.5$.

4. We need to find the best match of H to Li atomic orbital energies. Using the data that is provided, the H($1s$) energy of $-1312\,\text{kJ mole}^{-1}$ is closest to the Li($2s$) orbital energy of $-520\,\text{kJ mole}^{-1}$. The H($1s$) and the Li($2s$) orbitals will form a σ orbital and the Li($1s$) orbital will contain a non-bonding pair of electrons. The electron configuration will be $\text{Li}_{1s}\sigma^2$. Note that the bonding orbital has no designation since it arises from different orbitals on the two atoms.

Chapter 11

1. The Be electron configuration is $1s^22s^22p$. In our thought model, we first unpair the $2s$ electrons so that the electron configuration becomes $1s^22s2p^2$ and then hybridize the three $n = 2$ orbitals: $1s^2(sp^2)^3$. Be then forms three σ bonds to the three hydrogen atoms. The sigma bonds are formally composed of H($1s$) + Be(sp^2) to give a final trigonal geometry.

2. Looking at the butadiene and ethylene structure, we can see that butadiene is effectively a dimer of ethylene, so we expect the bonding to be related. Each carbon atom in butadiene is bonded to three atoms. Recalling that we treat a multiple bond as one "bonding partner", the expectation is that each carbon atom forms sigma bonds using sp^2 hybrid orbitals. If we look at the terminal carbon atoms, each is bound to

two hydrogen atoms and one carbon atom. The hydrogen bonds are formed by combining $2 \times \text{H}(1s) + \text{C}(sp^2)$ and the carbon-carbon σ bond is formed from $\text{C}(sp^2) + \text{C}(sp^2)$. The internal carbon atom is bound to one hydrogen atom and two carbon atoms. The bonds are formed from the same atomic orbitals: $2 \times \text{C}(sp^2) + \text{C}(sp^2)$ and $\text{H}(1s) + \text{C}(sp^2)$. In both sets of carbon atoms, the expected geometry about the atom is trigonal (120° bond angle).

4. We have just determined in Exercise 3 that the carbon atoms form bonds using sp^2 orbitals with a preferred geometry of 120°. The drawing for the cyclic form of butadiene has C–C–C bond angles of 90°. This molecule will be very strained and rapidly ring open to the acyclic form.

5. In MO theory, the O_2 electron configuration is given as $\sigma_{1s}^2 \sigma_{1s}^{*2} \sigma_{2s}^2 \sigma_{2s}^{*2} \sigma_{2p_z}^2 \pi_{2p_x}^2 \pi_{2p_y}^2 \pi_{2p_x}^* \pi_{2p_y}^*$, which provides a total bond order of 2 with a σ bond order of 1 and a π bond order of 1. In valence bond theory, the oxygen atom has the electron configuration $1s^2 2s^2 2p^4$. We imagine promoting one of the $2s$ electrons to the $2p$ orbitals to result in the $1s^2 2s 2p^5$ electron configuration. We know that O_2 is linear, so we expect that the hybridized orbitals will be sp. However, the oxygen atoms in the O_2 molecule each have two non-bonding pairs of electrons. Therefore, there are three bonding partners and the hybridized electron configuration is $1s^2 (sp^2)^5 2p_x$. The oxygen molecule bonding has a σ bond from $\text{O}(sp^2) + \text{O}(sp^2)$ and a π bond from $\text{O}(2p_x) + \text{O}(2p_x)$. The two models provide the exact same structure and bonding.

Chapter 12

1. The predominant force between all of these molecules, except for NH_3, is dispersion. For ammonia, hydrogen bonding

is possible because nitrogen is highly electronegative and the hydrogen atom from one molecule can closely approach another. NH_3 has the highest boiling point.

2. Polarizability increases with molecular size and linearity. CH_4 is essentially spherical and the linear hydrocarbons are increasing in length. Therefore, the longest molecule, C_4H_{10}, will have the greatest polarizability, the strongest intermolecular forces and the highest boiling point.

3. The forces present are: CH_4: dispersion; HF: hydrogen bonding; NO: dipole-dipole interaction; and C_2H_6: dispersion. Therefore, HF will have the strongest intermolecular forces.

4. Looking at Figure 12.4, the equilibrium position will be the minimum energy position, which occurs near 1 Å.

5. HCl is a polar molecule and will, therefore, experience attraction to the glass walls of the tube and yield a concave meniscus, much as does water due to these adhesive forces. Gallium is non-polar and will not have any interaction with the glass walls. Cohesive forces dominate and the meniscus will be convex.

Chapter 13

1. 1,3-butadiene has four π electrons (in the "box") and four carbon-carbon bonds. The total box length is $3 \times 0.140\,\text{nm}$ or $0.42\,\text{nm}$. The ground state will have two electrons in the $n = 1$ and $n = 2$ orbitals. From Equation (13.3) we know that the energy difference between the ground state and the first excited state (two electrons in $n = 1$, one electron in $n = 2$ and one electron in $n = 3$) is given by

$$\Delta E = E_{n+1} - E_n = \frac{(2n+1)\hbar^2}{8mL^2}$$

where the quantum numbers are $n = 2$ and $n + 1 = 3$.

$$\Delta E = E_3 - E_2 = \frac{(5)(6.62 \times 10^{-34}\,\mathrm{J\,s})^2}{32\pi^2(9.11 \times 10^{-31}\,\mathrm{kg})(0.42 \times 10^{-9}\,\mathrm{m})^2}$$
$$= 3.45 \times 10^{-19}\,\mathrm{J} = \frac{hc}{\lambda}$$

Solving for the wavelength yields 575 nm, in error by more than a factor of two. The issue here is related to the cyanine wavelength issue: the length of the box is incorrect. In this case, because the geometry around each carbon atom is trigonal, the box is actually shorter than our calculation, but again, this is a good exercise to understand some of the quantum mechanics models. A box length closer to 0.25 nm will provide an estimate in better agreement with the experimental value.

2. The carbon atoms in the cyclopropane molecule are each bound to four other atoms, and we would expect that the preferred geometry around each atom would be tetrahedral with 109° angles. The equilateral triangle formed by the carbon atoms appears to have 60° angles, so we would expect that this molecule would not be stable and would prefer to ring open.

Chapter S1

1. The boundary condition is that $f_1(x) = 0$ at $x = 0$ and $x = 1$. Equation (S1.8) defines $f_1(x) = \sin\sqrt{|C|}x$. At $x = 0$, $f_1(0) = \sin(0)$ and $\sin(0) = 0$. At $x = 1$, $f_1(1) = \sin\sqrt{|C|}$, which will satisfy the boundary condition if C is an integer multiple of π. Equation (S1.9) defines a function $f_2(x) = e^{(-\sqrt{C}x)}$. Now, $f_2(0) = e^0 = 1$. Hence, this function must be discarded.

Index

www.ingramcontent.com/pod-product-compliance
Lightning Source LLC
LaVergne TN
LVHW050934210926
848482LV00013B/615

* 9 7 8 9 8 1 9 8 3 1 7 8 4 *